The
Square Root
of Einstein
The Mysterious Connections
in Our Universe

The
Square Root
of Einstein
The Mysterious Connections
in Our Universe

Chris D. White
Queen Mary University of London, UK

World Scientific

NEW JERSEY · LONDON · SINGAPORE · BEIJING · SHANGHAI · HONG KONG · TAIPEI · CHENNAI · TOKYO

Published by

World Scientific Publishing Europe Ltd.

57 Shelton Street, Covent Garden, London WC2H 9HE

Head office: 5 Toh Tuck Link, Singapore 596224

USA office: 27 Warren Street, Suite 401-402, Hackensack, NJ 07601

Library of Congress Cataloging-in-Publication Data
Names: White, Chris D., author.
Title: The square root of Einstein : the mysterious connections in our universe /
    Chris D. White, Queen Mary University of London, UK.
Description: New Jersey : World Scientific, [2024] | Includes index.
Identifiers: LCCN 2024000570 | ISBN 9781800615427 (hardcover) |
    ISBN 9781800615526 (paperback) | ISBN 9781800615434 (ebook) |
    ISBN 9781800615441 (ebook other)
Subjects: LCSH: Physics--Popular works. | Physics--Philosophy--Popular works.
Classification: LCC QC24.5 .W495 2024 | DDC 530--dc23/eng/20240226
LC record available at https://lccn.loc.gov/2024000570

British Library Cataloguing-in-Publication Data
A catalogue record for this book is available from the British Library.

For any available supplementary material, please visit
https://www.worldscientific.com/worldscibooks/10.1142/Q0456#t=suppl

Desk Editors: Nambirajan Karuppiah/Shi Ying Koe

Typeset by Stallion Press
Email: enquiries@stallionpress.com

*Once upon a time, David "Chalky" White
and Jean Hammett gave their young son
a book about science.*

He now returns the favour.

Preface

This is the third volume of a trilogy, written for different audiences, and in the wrong order. The first book in the series was a textbook on the theory of electromagnetism (of which much more later), designed for first-year university students. This was a subject that I myself had struggled with in my youth, and I had always intended to write a book summarising the theory as I saw it and making connections to some exciting current research that I happened to be involved in. That research seemed to be telling us that our various theories of physics were much more closely related than we previously thought they were, spawning the second volume in my series: an advanced textbook that teaches this new research area to trained physicists so that they can become involved themselves. While thinking of what should go into that book, my thoughts turned back to the "popular science" books that I read as a teenager, and that got me into science in the first place. If my aim was to get more people to join my particular favourite corner of theoretical physics, why shouldn't I be starting earlier and trying to get people interested in science at a much younger age? It is also true that the general public likes to be kept abreast of scientific developments, albeit explained in a way that does not involve seven or more years of prior technical training. Thus, there seemed to me to be scope for such a book, exploring interesting recent developments in our current understanding of the theories underlying all of physics. The result is the book you are holding in your hands.

I shall give the standard preemptive apologies common to all prefaces of this nature. Opinions expressed throughout are my own, and if I happen to have focused on certain developments rather than others, this was to make my chosen narrative more straightforward and should not be taken as a comment on the work of others. I have tried to strike a balance between presenting lively scientific ideas on the one hand, and merely giving a list of names of who did what. This is not a history book, and the development of science as presented here follows my own way of putting the pieces together, rather than a strict chronological timeline. Pedants will find many places in which I have been somewhat loose with scientific terminology, in order to make meanings (as I understand them) leap out more clearly.

I am very grateful to Laurent Chaminade at World Scientific for initiating this project, also to Nambirajan Karuppiah, Ana Ovey and Shi Ying for their amazing help in the production process.

This book is dedicated to my parents. They are not scientists themselves, nor did they ever have a chance to go to university. Precisely none of this mattered for my scientific development, for they gave love and support in abundance and demonstrated a phenomenal work ethic which I spend every day trying to live up to. Most importantly, they gave me the single most useful gift that you can give to any budding scientist: *the ability to be constantly amused by the sheer daftness of the world around us!*

About the Author

Dr. Chris D. White is a Reader in Theoretical Physics at Queen Mary University of London. Originally from Cornwall, he studied at the University of Cambridge, obtaining a PhD on the structure of the proton, before holding positions in Amsterdam, Durham and Glasgow. Chris has published over 70 research papers on a wide range of topics in high energy physics, ranging from practical calculations for particle accelerators such as the Large Hadron Collider, to relationships between particle physics and gravity. He has also won teaching awards recognising his efforts to make physics accessible to students from underrepresented backgrounds.

Contents

Prologue

It is possible to arrive at the Higgs Centre for Theoretical Physics – located at the University of Edinburgh in the United Kingdom – in less than an ideal mood. The buses from the main station always seem just unreliable enough to justify walking the 40 min or so to the institute. Upon doing so, it will inevitably start raining, such that you will decide to duck into one of Edinburgh's many tourist shops to buy a cheap umbrella. You will quickly discover that the umbrella is useless: strong winds will blow horizontal sheets of rain directly into your face, soaking your clothes in the process. The gusts will then destroy the umbrella.

None of this turns out to matter: upon arriving at the Higgs Centre, your cold and soggy exterior will be warmed by some of the friendliest scientists you are ever likely to meet and secretaries who are able to organise your life much better than you are. It is thus very common that there are one or more scientific visitors at the Higgs Centre at any time. Something more unusual, however, happened in 2014, when about 30 visitors turned up all at once. They had been invited to a *workshop*, a name commonly used by scientists for a meeting in which people tell each other what they are working on, before seeing how this knowledge can be fused together to create new scientific results. Attending a scientific meeting can be quite terrifying, especially so if you are a scientist! If you are not, the nearest thing I can compare it to is attending parties when you are a teenager, despite a few obvious differences (e.g. at a physics conference, the nerdiest people often turn out to be some of the most popular!). First of all, there is the fear that you will not be invited at all and will miss out on all the fun. But the relief at being granted an invitation soon gives way to yet more fears, regarding what you will say to people, whether they will pay attention, and whether they will speak to you afterwards.

I did not have to worry about being invited to the meeting in 2014, as I was one of the organisers.[1] To stretch the above analogy, this is rather like holding a party in your own house. Can you really enjoy yourself, if you are constantly having to make sure other people are enjoying themselves and not damaging the furniture? It was indeed a somewhat stressful week. As with many a UK academic conference, organisation was occasionally informal, with trips to the local supermarket being substituted for more fancy catering, and lecture rooms functioning as makeshift kitchens where needed. None of the participants seemed to notice this, possibly due to the copious amounts of local whisky that we offered them. There was strong coffee too, the first cup of which was offered to an American professor who arrived on the first morning straight from the airport, having flown direct from Los Angeles.

As the week went by, people mostly found themselves congregating in the Higgs Centre's *seminar room*, a large carpeted space with sofas and a kettle at one end and six large sliding blackboards at the other. There are chairs and tables laid out so that people can pay attention to anyone who might be presenting their work at the blackboards and make notes. Covering the long side wall of the room is a further set of blackboards so that when talks are not in progress, many groups of people can discuss at the same time and listen to each other's conversations so that they can pitch in ideas. By the end of the workshop, all of these blackboards were completely covered in diagrams, algebra, and mysterious words, not unlike the representations of "physics" that you see in Hollywood films. To an insider, though, the mysterious hieroglyphs meant something, at least for odd moments in the few minutes during which each discussion took place. By the end of the week, people were so enthused by what had gone on, that it was decided to hold the meeting again the following year, but this time in Los Angeles.

---

[1]My fellow organiser was Donal O'Connell, based at the University of Edinburgh.

There must, of course, be a reason why a group of 30 people flew half way around the world to talk to each other for an entire week and an even better reason for their deciding to keep at it. But the trouble with science nowadays (and perhaps ever) is that I cannot simply tell you what it was that was so exciting, in language that the scientists themselves were using. It would take you literally years to reach the stage at which some of the symbols on the board started to make sense. And understanding only some of the symbols is the best you could ever hope for anyway: one of the great mistakes I made as a younger scientist is to think that a group of older scientists waffling on with chalk in their hands actually knew what they were talking about. In fact, science is now so specialised that most people in any conversation are desperately trying to keep up! I remember this feeling well that week, and in one particularly heavy discussion, I heard a phrase which for me came to sum up the meeting. "No that's not right", exclaimed an enthusiastic physicist who almost spilled their coffee, "because *the square root of Einstein is Yang–Mills!*". In the context of our discussion, and with the particular symbols that formed a backdrop behind us on the board, this statement made perfect sense. But I could not help drifting off slightly in my mind and reverting back to my own teenage years, in which I organised the rather terrible parties alluded to above. Unlike many teenagers, I was keenly interested in science, and physics in particular. But even given that, if you had used a phrase like "the square root of Einstein" to me back then, I would have thought you were talking the most ridiculous nonsense I had ever heard in my life. What can it possibly mean? And why on Earth would it drive 30 otherwise normal people to organise regular meetings, like some sort of newly formed religious cult, whose rituals include coffee-drinking and writing odd symbols on walls?

This observation hit me so forcefully that I have chosen to use this strange phrase as the title of this book. But trust me, "the square root of Einstein" is a phrase that encapsulates a great deal of profound mystery and that touches every single aspect of the universe that we currently live in. It turns out, you see, that all of the theories that we currently use to describe the universe are much *much* more closely related than we ever thought possible. So much so that it currently seems that we have been writing our theories of physics for the past few hundred years in the wrong (mathematical) language so that similarities between widely different areas of our experience have remained hidden from us. These previously hidden connections relate the smallest things we know about (the tiny particles that live inside the centres of atoms) to the very largest (black holes, galaxies, and the entire universe itself!). It is this idea that has gotten so many scientists highly excited, and the aim of this book is to explain this to you in terms that are understandable without first having to spend several years finding out what various funny symbols mean.

We already have a clear aim in mind: can we at least get as far as understanding the title of this book? Indeed we can, but there are a large number of other things that I have to tell you first. Before doing this, it is only fair that I spend the following chapter explaining why you should care about what I have to tell you.

Chapter 1

Twinkle Twinkle

Science, especially physics, has an image problem. Casting my own mind back to when I was at high school, an interest in physics or maths was a perilous activity and liable to cause you to be called a *nerd, geek, dweeb*, or *square*. Those were the politest words, and this was after all thirty years ago. I am confident that much stronger words will exist nowadays, and the above words will now have gone sufficiently out of fashion that a young person using any one of them will themselves be marked out as whatever passes for a square these days. However, there is something puzzling about this situation: speak to any young child about anything for more than two minutes, and the question you will hear most often from them is "Why?". Usually this takes the form of immediate concerns, such as "Why can't I have a biscuit?", but it is undeniable that children provide very good evidence that we are a naturally curious species that wants to establish its place in the world. They don't simply accept things but demand to know why they should care about and agree with them. Children learn by constantly probing the limits of what they know already and – to the amusement or horror of adults – often have no preconceptions for how an unknown situation will affect them. They are singularly fascinated by where things come from, be they people, animals, or mountains, and also why things happen the way they do. Almost all adults agree that this is a very good thing that should be positively encouraged. In the UK at least, many parents sing nursery songs to their children to encourage creativity and language skills, one of the most popular being *Twinkle twinkle little star, how I wonder what you are!* How can any child not respond to such a musical plea to understand the wonders of our universe?

Unlike many parents who sing this song, I am well aware of what stars are made of and what it is that makes them twinkle. But I am also aware that if you try to tell someone that *physics* tells us how this works, you might face anything from their eyes glazing over to outright hostility. Physics and maths, it seems, leave mental scars that remain with people for the rest of their lives. They are referred to as the "difficult" subjects at best or the "useless" subjects at worst. At all levels in our society – including on flagship BBC news programmes and at dinner parties populated by university graduates – it is perfectly acceptable to claim ignorance of science and maths and even to revel in this lack of knowledge. There is a camaraderie in doing so, even for those snobbier characters for whom ignorance of literature, history, or politics is looked down upon as a sign of intellectual weakness. And yet these same people are happily singing about twinkling stars to their children!

Another contradiction happens when you speak to many adults. Despite saying that physics and maths are not for them, they remain genuinely interested in the big questions in life. What are we made of? What is our place in the universe? What is the universe as a whole made of? Was there a beginning, and if so, how did the universe get here? Will there be an end? Why is the universe the way it is, and could it have been different? These questions are the remnants of our once highly active childhood curiosity, and it is easy for us to end up dismissing them as either unanswerable or impractical. Even if we don't, as we get older, the sheer business and exhaustion of our daily lives make it difficult to even remember the big questions we once contemplated, let alone answer them. But – to a partial but remarkable extent – these questions *are* answerable, and it is *physics* that answers them. How can it possibly be true that someone who is interested in the big questions can simultaneously think that physics is "not for them"? When people think of our collective culture as a species, it is usually art, music, literature, theatre, and/or cinema that come to mind. Why is science not only absent but also thought to be impossible to add to the list of things that all people are able to be interested in?

Of course, there is a small group of people in society who never lost sight of the big questions mentioned above and who continue to try to answer them. They are called *physicists*, and they are constantly trying to find new ways of thinking about the universe or to test these ideas by doing experiments. This is what we call *fundamental physics*, which roughly speaking means understanding the universe for its own sake. There is also *applied physics*, in which we use existing theories to develop new technologies, that enrich our lives and create economic prosperity. There is no hard boundary between these two disciplines, and you will very often see the same people working in both areas. Furthermore, applied physics clearly links to other areas of science and technology, such as engineering, chemistry, biology, medicine, computer science, and more. Roughly speaking, what a physicist does is to write laws governing how the universe works that allow us to predict what will happen in future. It is these laws that have given us astonishing answers to the big questions we saw above, and the correct language for these laws is that of mathematics, which can often be very advanced. The reason mathematics occurs in physics is really quite simple: everything that an equation says can actually be written in words if we wanted to. But we would need huge numbers of words, and we would also end up repeating ourselves a lot, due to the fact that certain very complicated ideas end up occurring again and again in many different places. Mathematics gives us a shorthand that is extremely efficient and able to be understood by multiple people across different cultures. Despite this mathematical complexity, however, the basic ideas behind the laws can be appreciated by anyone, regardless of their mathematical background.

So how did we get where we are? How can it be true that people start out in life by being highly interested in how the world works – in many cases retaining this curiosity in adulthood – but end up deeply phobic about the very subject that can tell them the answers? I have my own thoughts about this, but let me first explain why it is such a problem. As we see in this book, fundamental physics is at a turning point. Our theories to date have been extraordinarily successful and have allowed us to describe the universe from mere fractions of a second after its beginning to the present day. However, some very significant holes remain in our understanding, such that new theories are needed. Furthermore, as mentioned in the prologue, it seems that the mathematical language we have been using up to now may itself have misled us, such that we are getting tantalising glimpses of an incredible underlying structure behind everything we know. We are also, as a society, facing extremely complex problems, including how to guarantee our collective health in the coming decades, as well as the health of our planet. It is applied physics that can provide new solutions to these problems, and thus – from either an esoteric or a practical point of view – it has never been more important for young people to pursue careers in science. In a way that is genuinely different to the problems encountered by previous generations, it is not at all obvious where the next big breakthrough will come from, and people from all walks of life, regardless of gender or cultural background, could potentially be the scientific leaders of the future. This can only happen if they view a scientific career as an actual possibility in life and are encouraged from an early age to work towards it.

Unfortunately, this doesn't seem to be happening. The society we live in is very much one in which something happens in later childhood or adolescence, that somehow takes all the curious would-be scientists and convinces them not to bother. That's fine if people genuinely want to do something else for a living, but many of these people end up regretful later in life, particularly after they have encountered a younger generation, asking the very same questions that they once asked of their own parents. Entire volumes could be filled with the research papers trying to understand what happens and, more importantly, what should be done about it. Often, however, it comes down to some simple perceptions, many of which turn out to be myths. Let us examine some of them here:

(1) *You have to be a genius to be a scientist*: Ask many people to describe what a scientist does and they will probably think of a lone genius, who sits in a chair and thinks all day. Very occasionally, they will scribble some algebra on a piece of paper and be awarded a Nobel Prize for their efforts. Even without this stereotype, the idea that you have to be somehow incredibly talented or special to do well in science is widespread, and media depictions of scientists are largely to blame for this. The actual reality of being a scientist is almost the complete opposite! Walk into any research lab, including those that only do calculations without any experiments, and you will see groups of people working together. Collaborations also happen remotely, proceeding via email and video calls. As in any job, when people work in a team,

they bring different skills and techniques to the table and learn from each other such that the team is greater than the sum of its parts. Why isn't this the image of science that we are given in childhood? It means that anyone can be a scientist and that you do not need to somehow be recognised as a prodigy or chosen one in early life!

(2) *You have to be good at maths to be a physicist*: This is true, but doesn't have to be a problem for anyone. Much like the above point, anyone can be good at maths, provided they practice and learn from other people. The rewards are very high indeed: as we will explain at multiple points in this book, the laws of physics are written in mathematics. Not only this, but they also condense our combined experience to such a powerful extent that almost everything we see in our daily lives can be seen to follow from an incredibly small set of underlying principles. This is truly amazing!

(3) *Maths and physics are not creative subjects*: A common reason for people abandoning science for other subjects early on is that the arts (e.g. literature, drama, and music) or humanities (geography and history) are seen to be much more creative than maths or physics, which always seem to involve a "right" or "wrong" answer. Depending on how the subject is taught, students will come away feeling like failures if they get the "wrong" answer. Even if they get the "right" one, they will wonder why they bothered, if there was no freedom to get anything else. This is *completely unrepresentative* of what research in science is like! In research, we are always trying to generate new knowledge and discover things that nobody has noticed yet. We do this by working together and building on everything that has come before. The theories that we talk about in this book emerged after some of the most creative acts that the human species has ever performed. Listing the achievements of some of the individual people involved does no justice whatsoever to the billions of creative acts that happened over thousands of years, in every continent and culture on Earth, and that somehow ended up being distilled into our current theories of the universe. As a working scientist, you use your creativity on a daily basis, playing with ideas and methods in order to develop new ones. You bounce off other people, both teaching and learning from them at the same time. You are

sometimes amazed that different things you thought you knew turn out to be the same thing, and the experience of combining scientific ideas feels very much like when you played with your childhood toys.

(4) *Maths and physics are not beautiful subjects*: Even if science is creative, some will concede, it still doesn't make it beautiful, in the way that poems and paintings can be. This is an alarmingly common misconception, born from ignorance. Take it from me – you cannot encounter the theories we describe in this book without being dumbstruck by their sheer beauty and elegance. Art has the power to tell us great truths about our existence and our relationship with others. But our best scientific theories, as well as being standalone mathematical jewels in themselves, are also incredibly powerful in their ability to predict how our universe behaves and supremely economical in how they achieve this. They are the most extreme understatements ever uttered! I do not mean to imply that things are perfect: our current theories have their fare share of ugly bits and warts. But that is why we need more scientists.

(5) *You cannot be a scientist if you are religious*: This is a very common myth so needs to be called out. I can tell you from personal experience that many of my research collaborators are deeply religious people, and examples of famous scientists of any religion are easy to come by. Indeed, religion often seems to act as a motivating factor for some scientists, who regard their physics research as a very close scrutiny of their god's (or gods') creation. The breathtaking beauty and simplicity of some of our contemporary scientific theories certainly adds to this sentiment, and there is clearly no contradiction in being a religious scientist. Even atheists usually refer to the mythic "Nature" as being somehow responsible for the laws we discover.

(6) *Physics and maths are impossible*: Not true – they are merely very difficult, which was nature's choice. If you choose to study them, though, you will never be alone, and you will have many people to help and guide you.

Ultimately, the main reason people turn away from science in their teenage years might simply be that this is precisely the time at which we are most concerned by what other people think of us. If liking maths and physics makes you a nerd, why would you choose to be associated with these subjects when you are most likely to suffer as a consequence? My only suggested remedy for this is to imagine a world in which science and maths *are* cool and are widely recognised to be the highly creative and awe-inspiring subjects that they genuinely are. Imagine people chatting about the latest scientific discoveries like they chat about the latest hit songs or films or controversial politics. People from many different backgrounds would choose to study science, unencumbered by being seen as some sort of misfit. This in turn would lead yet more people to want to join and to a much more diverse and efficient scientific workforce than the one we see today. It is in all our hands to create this society, and anyone can help. My aim in this book is to equip you with the tools to help, should you wish to do so. I will tell you the basic ideas behind all of our current scientific theories, before showing you what I think is amazing evidence that there is something mysterious going on behind them. Are you a young person who is scared of pursuing science or think you are not able to do so? Please use this book to develop the main weapon that you will need – enthusiasm – to go further and to allow yourself to be curious. Please also read the last chapter, in which I explain more to you about what a scientific career actually involves. Are you a parent of a curious child? Please use this book to make sure their curiosity is never extinguished! Are you neither of these, but simply want to find out more about what scientists are currently up to? I am at your service, but please do what one of my university lecturers once asked of me: go forth and spread the message of science! Let people in on the secret that it's nothing to be scared of but everything to be amazed by!

Chapter 2

The Most Obvious Things Aren't True

Having hopefully succeeded in convincing you why physics, and the love of physics, is important, I now want to start to show you all of the theories of physics that we currently have. I also want to get across how all of our theories fit together, or don't, as this will lead on to discussing quite why our understanding of this has changed so dramatically in the last few years. Unfortunately, I cannot simply show you the complete set of what we call "physics" right away – it would be unfamiliar, nonsensical, and downright scary! Instead, I follow something like what a formal education in physics does. That is, I start with things that feel familiar to us from our everyday life and gradually work outwards from there until we can understand everything that scientists themselves currently understand. These theories describe everything from the very weird goings on inside atoms to where the Big Bang comes from. By comparison, the physics of our everyday lives can seem boring and unglamorous, and many beginning physics students (as I did back in the day) find themselves skimming over this to get to the good stuff. They do so at their peril, as the physics describing the everyday objects we see around us already contains many of the concepts that are needed for understanding the more exotic things mentioned above. I give you the key ideas in this chapter.

To start us off, let me ask you to look around you and describe to yourself what you see. I am currently writing this in my office in

East London, for example, and my desk is that of a typical physicist: laden with books, papers, notebooks full of random algebra, and coffee cups. There are borders to my office that I cannot see through ("walls") and those that I can indeed see through ("windows", albeit those in need of a clean). Clearly, all of these things are made of something, and thus the conclusion we reach is that the universe contains an awful lot of *stuff*. Also, not all of this stuff is in the same *state*. There are solid things like the books and then liquids, such as ink or coffee. There is also stuff that I cannot see but can feel the effect of, such as the air – a gas – that I am breathing in and out. If I look out of the window, I see an enormous variety of other stuff, some of which is moving. If you are looking around your environment while reading this, you will soon lose count of all of the different types of things, materials, colours, textures, etc. that you spot, such that it feels completely impossible that all of these things could be at all related. Let us not worry about this for now and simply note that there are things in the world and that they must be made of something. Physicists use the word *matter* to describe all the stuff, and it is clearly something that our theories of physics have to describe.

Now let us do another experiment. First of all, hold something that is not too fragile (or that you don't mind damaging) at arm's length and let it go. You know already what will happen from your experience of course: it will fall to the ground, speeding up as it does so. It will then hit the ground and possibly bounce. There is clearly something making the object move, and this is called a *force*. Here there are actually two forces involved – the one that makes the object fall to the ground and the one that makes it bounce. You probably know what the first force is (gravity), but you may not know what the second one is. If so, please don't worry, as we will get to this later on. And there are anyway many more forces than just these. As you go about your daily life, you can pull, push, or kick things to make them move. You can swing things about on ropes, blow out candles, drive cars, bake cakes that rise up in the oven, and throw snowballs. You can splash water about in sinks, stroke the fur on pets, communicate with other people using phones and computers, and ride a bicycle. All of these things involve something moving, whether or not you can see it directly. Furthermore, even if things are *not* moving, there may be forces involved: if you sit in an armchair, you are not pulled through the surface of the Earth. Gravity wants to do this, but the chair somehow creates a force that cancels this out so that nothing happens. However, it remains true that the universe contains *matter* (the "stuff" alluded to above) that is acted on by *forces*.

The distinction between matter and forces is perhaps the most basic distinction we can make in all of physics, and it also gives us a clue about how to understand the world around us. First, we can classify all the matter that things are made of. Second, we can understand all of the forces and describe how these act on matter to create the changes in behaviour that we would like to be able to predict. The branch of physics that describes how objects are moving is called *mechanics*, and the laws that apply to objects in everyday life are usually referred to as *Newtonian mechanics*.

To say more about what this means in practice, we need to be able to talk about how objects move, which I would normally do using algebra. However, the maths is really just a shorthand for a lot of words so that we can instead use words if we like, provided we are suitably precise. The first concept I want to introduce is the *speed* of an object, which corresponds to how fast something is moving. You will be very familiar with this from your everyday life. If you look at an average street, you will see lots of people and vehicles moving at different speeds. If you drive a car, you will even see a little dial or digital meter that tells you *how fast* you are moving, in miles or kilometers per hour. How speed is measured tells us what it means: if you are moving at 20 mph, you would cover 20 miles of distance within an hour if you carried on travelling at the same speed. Thus, speed conveys a notion of how much distance you can go in a certain amount of time. We can use any measure for the distance (e.g. miles and kilometers) and any measure for the time (e.g. hours, minutes, or seconds), provided we also quote these measures when we give a speed.

Speed is not the only quantity that is useful for saying how objects move: they can also move *in a particular direction*. If you are about to cross the road, and I tell you that a car is moving near you at 30 mph, you will want to know if it is moving *towards* or *away* from you. Only in the former case is there a problem, and this tells us that the speed is not enough to fully describe how objects are moving. Physicists deal with this by defining something called *velocity*, which tells us how fast an object is moving *and* which direction it is moving in. As a word, *velocity* is used less in everyday life than *speed*, and some older readers may already encounter twinges of school-related panic upon hearing it. There is something somehow intrinsically nerdy about the word *velocity*, possibly due to the fact that Hollywood films can – and often do – instantly make someone sound like a boffin by having them speak it. Ultimately though, having a word which means both speed and its direction is very useful, as it allows us to be very precise when we talk about the motion of objects. For example, "the wind is blowing at 70 kilometers per hour due east" is a statement about velocity, but "the boat is moving at twenty miles an hour" is not (there is a speed, with no direction). Likewise, if I say that my cat is going due north, I have also not told you her velocity, as you do not know her speed.

Later on, we will need to know about a certain property regarding velocity, which you may have noticed already from your experience. If a stationary object is hit by a moving one, this usually causes some damage. However, if both objects are moving *towards each other* when they collide, it does even more damage! Imagine that you are in a car, for example. If you are travelling at 20 mph towards a second car, which is also moving at 30 mph towards you, you would effectively see that car as moving towards you *faster* than 30 mph. This is called *relative velocity*, and you may or may not know that the second car effectively moves at 50 mph from your point of view, the speed being given by adding the two other speeds together. This is not necessarily obvious, so do not worry if you don't happen to know it (I will say where it comes from below). I have also drawn a picture to summarise this concept in Figure 2.1, in case it helps.

Objects do not typically travel at the same speed for very long. Nor do they keep going in the same direction. If you don't believe me, go to the park and watch some dogs or, even better, some flying birds! These facts make it useful to introduce another concept that

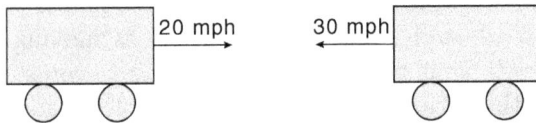

Fig. 2.1.   A car travelling at 20 mph moves towards another moving at 30 mph towards the first one. If you are sat in the first car, the second one seems to be coming towards you at 50 mph, a property known as *relative velocity*.

describes how velocity is changing. This is called *acceleration*, and there are broadly speaking two ways that objects can accelerate, given that velocity corresponds to speed and direction: (i) an object does not change direction but speeds up or slows down (the latter is often called *deceleration* in everyday language, but a physicist would use *acceleration* for both effects, given that they are both changes in speed and therefore velocity) and (ii) an object does not change its speed but does change its direction. For an example of the latter, think of a car going around a roundabout with a constant speed. Its direction is certainly changing, and thus it is accelerating. What makes this very confusing for beginning physics students is that we are so used to saying that acceleration implies something is speeding up. This, however, is a confusion that arises from a widespread misuse of words. Once we *define* acceleration to relate to a change in velocity, an object can have a constant speed but still be accelerating! We then have a single concept ("acceleration") that covers many different situations.

An object with a large acceleration has a velocity that changes quickly. Likewise, an object with a small acceleration has a velocity that changes slowly. The extreme case of zero acceleration means the velocity is not changing at all so that an object will keep going in the same direction, with the same speed. Armed with these various concepts, I can now describe what the basic laws of Newtonian mechanics say. Any object has an acceleration (which may of course be zero), and it then turns out that this is directly related to the overall force that acts on the object. This is usually called *Newton's (second) law of motion* and was first written down in the 1600s, based on careful experiments by many other people. Put very simply, *forces cause changes in velocity*. Thus, this law of physics makes very precise the idea that forces act on matter, to tell it how to move! It also cleared up some long-standing confusions among philosophers and their ilk. It used to be thought, for example, that forces were needed to *keep objects moving*, as in our everyday life we often see moving objects apparently slow down and stop of their own accord. However, this is actually due to forces that we cannot see directly, such as friction or air resistance. Were these forces absent (e.g. driving on an ice rink rather than a road or in a vacuum instead of air), an object subject to no force would simply keep going, as there is no force to create an acceleration, and thus no change in the object's velocity.

We have gotten as far as saying that forces cause objects to accelerate. But it is easy to see that not all objects behave in the same way: try throwing a tennis ball at someone, followed by a pumpkin. Even if you throw the pumpkin just as hard as the tennis ball (i.e. you "apply the same force"), the tennis ball will go much faster and travel further as a result. In other words, the pumpkin accelerates less and thus appears to somehow resist being moved, more than the tennis ball does. We have here talked about stationary objects being thrown, but the same principle applies to moving objects. If I were to send a Jack Russell and an elephant towards you at the same speed, which would you rather try to deflect? Your own proclivities are up to you, but there is no doubting that the elephant is more cumbersome and less easy to get to change its course. It seems, then, that all objects have some sort of innate property that describes how they resist changes in their motion. This is called their *mass*, and we then say that a more massive object accelerates less, for a given force. The laws of Newtonian mechanics tell us, in a precise mathematical way, how the acceleration on an object depends on the force applied and the mass of the object. The maths also allows us to rigorously derive things like the relative velocity property we mentioned earlier, whereby we add the speeds if we are moving towards something that is also moving towards us.

The earlier discussion is a bit dry, even for a seasoned physics enthusiast like me. It may therefore not be obvious quite how amazing the above ideas are! What Newtonian mechanics is saying is that *any* object, *anywhere in the universe*, obeys the same laws, provided it is suitably big and moving slowly enough. This would have been truly revolutionary when it was first proposed: if you go back several hundred years (in Western culture at least), then proclaiming the idea that objects in the "heavens" obey the same laws as those in our everyday world would have had you burned at the stake! Furthermore, provided we know how to describe the relevant forces, Newtonian mechanics allows us to calculate a stupendous range of phenomena. Sports scientists use it to improve the performance of athletes. Astronomers are able to predict the motion of the planets, stars, and even whole (clusters of) galaxies. Engineers can build stable bridges and buildings. Planes can fly, boats can float, bicycles can bike, and scooters can scoot. This is not to say that all situations are fully understood: nobody in the world, for example, currently understands how bumblebees are able to fly at all! But Newtonian mechanics allows us to understand most of what we see around us.

Before about 1900, many people believed that Newtonian mechanics was all there was to understanding the universe. All we had to do was to fill in the description of all possible forces, and also find out what all the matter was, and we would have a complete description of nature. Not only does this turn out to be false, but it also turns out to be so spectacularly wrong that we have to abandon many of the things we think are obviously true about the world we live in. If you look very carefully above, we assumed an enormous number of things in even talking about what a speed was. I very casually used the word "time", for example, and by using this word – as many of us do every day – we are implicitly assuming that time is the same for everyone. I don't literally mean the value of time on a clock: we have different *time zones* around the world, so if you live in Tokyo, you add a certain number of hours to the reading on the clock in my office in London. However, we tend to accept that our two different clocks would tick at the same rate: one second in Tokyo is the same as one second in London. This fact seems so obviously true that you probably have never even considered that you needed to assume it. Newtonian mechanics is built on this notion of *absolute time*, and here's the big shock: *it is not true*. Clocks do indeed tick at different rates for different people, depending on how fast they are moving, or if they happen to be near a very massive object. If you are confused by this, that is perfectly natural, as it goes against everything that feels sensible and "right" about our daily existence.

Another assumption we tend to make about time is that we all have a shared future and past. If I were to draw time increasing on a line as in Figure 2.2, I could mark a certain point and call it "now", or *the present*. Anything to the right of this I would call *the future*, and anything to the left I would call *the past*. Another way of saying that time is absolute is that we would all then share the same future, and the same past, and likewise the same "now". But I have just told you

Fig. 2.2. If time is absolute, we all share the same past and future. The present ("Now") divides these two things.

that in fact time is not absolute, and thus these basic notions of future and past are not true either! I don't expect you to understand any of this, nor to be able to contemplate what the implications are – we will see what they are in the following few chapters. First, however, I must explain to you how we know that Newtonian mechanics and its absolute time break down, and the answer appears when we study where forces come from ...

Summary

Before we move on, it is useful to recap the main ideas of this chapter in a short summary. At the risk of sounding like a lecturer (which I do of course happen to be), my hope is that this helps check your understanding of what has come before. It may also be useful to refer back to this summary when ideas from this chapter recur in later chapters. In this chapter, we have seen the following:

- The universe contains *matter* that is acted on by *forces*.
- Objects made of matter have a (possibly zero) *velocity*, which tells us how fast they are moving and in which direction.
- *Newtonian mechanics* tells us precisely how objects accelerate (change their velocity) in response to forces.
- Every object has a *mass*, where a higher mass means it is more resistant to changing its velocity.
- The equations of Newtonian mechanics tell us that when objects move towards each other, the velocity of one relative to the other can be found by adding together individual velocities.

Chapter 3

The Light of the Charge Brigade

In the previous chapter, we saw that the universe contains stuff called *matter*, and this is acted on by *forces*. In this chapter, we look in more detail at what the matter itself is made of. We then look at a particular type of force in detail, which leads us to the conclusion that the Newtonian mechanics introduced in the previous chapter does not in fact describe the world we live in.

3.1 What Is Matter Made of?

Imagine I buy an exotic cake from my local baker and give you this as a mid-morning treat, on the proviso that you correctly guess what is in it. You would probably start by bringing the cake closer to your eyes, such that you can look more closely at its structure. You might then start to see various pieces of fruit, chocolate, or other goodies inside the cake. Larger constituents such as cherries would be the easiest things to spot first, and you would then find yourself having to look a little closer before you could spot smaller constituents, such as raisins or chocolate chips. We can clearly perform a similar operation on any object we see around us, and it is natural to ask what happens if we zoom in ever closer.

You probably already know the answer to this. Most of what we notice around us is made of *atoms*, where these atoms are held together in various ways to make larger composite structures called things like *molecules*, *crystals*, *metals*, and so on. I do not worry much in this book about the larger structures that atoms can make.

Instead, let us focus on what makes up an atom itself. First, we can note that there are many different types of atoms, each of which will have a different size. However, the typical size of an atom is about 0.0000000001 m, which certainly explains why we don't "see" atoms with our own eyes. If we choose to zoom in further, we will find that an atom is not a single thing but instead contains smaller things inside it. Every atom has a relatively small central object called a *nucleus*, and there are other objects called *electrons* that go round this nucleus. We will not yet worry too much about how the electrons go around the nucleus, but the most naïve thing we can imagine is that they might orbit the nucleus in a similar way to how planets orbit the sun. This simple picture turns out to be false, in such a spectacularly bonkers way that we cannot yet do full justice to it! But let us indulge our naïvety for now, in which case we might draw the simplest possible atom, containing a single electron, as something like Figure 3.1(a).

In trying to understand matter, we have now shifted the problem to trying to work out what electrons and atomic nuclei are made of. Based on the above remarks, we should now try zooming in further to see what happens, and indeed something interesting occurs. All experiments that have ever tried to zoom in on an electron have found that it apparently has nothing "inside it". Put another way, the electron apparently has zero size and is what we call a *fundamental particle*. It is one of many fundamental particles in nature, and we may think of these as building blocks, out of which everything else in nature is made. This property of the electron – that it is a fundamental particle as far as we can tell – is reflected in our current theories of nature at its smallest distance scales, as we will see.

Now let us look at the nucleus. We know that it is much smaller – by about 100,000 times – than the size of the atom as a whole, where the latter is set by the distance at which the electrons can be found orbiting the nucleus. However, unlike the electron, the nucleus does indeed turn out to have other things inside it if we zoom in, as shown in Figure 3.1(b). It contains two types of objects, called protons and neutrons, which itself resolves the mystery of why there are different types of atoms: these simply correspond to having different numbers of protons and/or neutrons in the nucleus. The way a given atom can take part in chemical reactions is entirely determined by the number of protons, which also controls how many electrons can be found

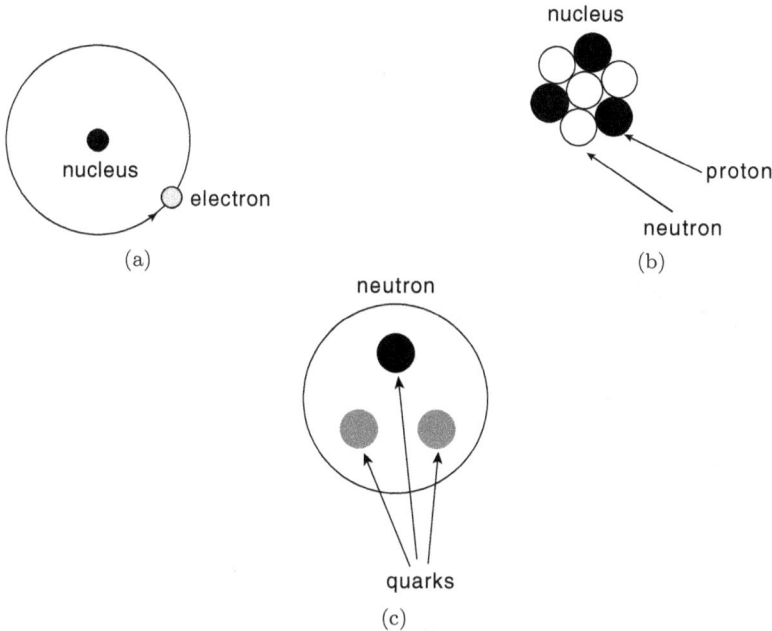

Fig. 3.1. (a) Atoms consist of a small nucleus, which is orbited by particles called electrons (one of which is shown here); (b) zooming in on the nucleus reveals the existence of protons and neutrons; (c) zooming in on a proton or neutron reveals particles called quarks living inside them.

orbiting the nucleus. Atoms with the same number of protons, but different numbers of neutrons, are called *isotopes* and have similar chemical properties. For those that are really interested, the number of protons in a given atom is called its *atomic number* and tells us where it sits in the famous *periodic table* that lists all the possible types of atoms that we have ever seen – or even made artificially!

Chemistry is a lovely subject but alas not the subject of this book. Let us thus zoom in further and ask if the protons and neutrons are fundamental particles. The answer is no, in that they contain further particles called *quarks*, as shown in Figure 3.1(c). Experiments tell us that there are six different types of quarks in our universe, and they have very unusual names, reflecting the somewhat strange history of the subject, as well as physicists' often misguided attempts at humour. The different types of quarks are called (in order of how massive they are, with the lightest first) *down, up, strange, charm, bottom*, and *top*. The proton and neutron (mostly) contain up and down quarks. We can see the additional quarks in particle accelerators such as the Large Hadron Collider at CERN or as the result of collisions of very high energy particles that enter the Earth's atmosphere from space (so-called *cosmic rays*).

As far as we know, zooming in further will not reveal any additional structure, such that the quarks themselves are believed to be fundamental particles. This then completes our survey of what is inside the atom, at least for now. Let us then ask: is this all we need to describe all matter in the universe? The answer is not quite. There turn out to be two other fundamental particles which are almost identical to the electron but have larger masses. Known as the *muon* and *tauon*, they can be seen in particle accelerators or cosmic rays, as earlier. There are also particles called *neutrinos* that can be produced in certain radioactive processes. There is one type of neutrino for each of the electron-like particles, and thus they are called the *electron, muon, and tauon neutrinos*, respectively. This completes the list of fundamental matter particles, such that we can very confidently say what the building blocks of matter are. We return to the theory that describes these in Chapter 6.

As we have seen, matter is only one side of our current description of nature. We must also explain where the forces come from, that act on matter to create interesting motion. Indeed, forces must be at play in our earlier description of the atom, in that there must be something that causes the electrons in the atoms to orbit the nucleus, just as the force of gravity causes planets to orbit the Sun. Let us now examine this in more detail.

3.2   Forces and Charges

In everyday life, forces can be observed every time we push or pull something. Likewise, when we throw something, we are very conscious of the effort required in throwing it, i.e. that we are exerting a force. After an object is thrown, we can clearly see the force of gravity acting on it, such that it gets pulled towards the surface of the Earth and hence the ground. Other examples of forces you may have experienced include the strange knot in your stomach if you are in a car or train that stops very suddenly, the jerks and twists you experience on a roller coaster, the force of the current in a river or sea that you are trying to swim in, or the force of wind against your face or clothes. In fact, we are bombarded by forces so continually in our daily – and nightly – lives that we tend to not even notice that this is happening. Even if we acknowledge this experience, it does not seem to help us in understanding what forces actually are or where they come from. Whereas matter is some sort of tangible stuff that we can touch and feel, forces instead correspond to some intangible part of our experience, corresponding to the touching and feeling itself.

Although our current theories of physics still maintain the distinction between matter and forces, they do so in a way that places them on a remarkably similar footing that is totally at odds with our direct perceptions of the world. In order to begin to understand this, let us focus on one of the particular forces that is involved in holding atoms together. That is, we describe precisely what is happening to make the electrons orbit the nucleus. We can start off by noting that we were very vague in talking about electrons earlier. We noted that these were fundamental particles and stated that these have a different mass to other related particles (the muon and tauon). However, we did not completely specify all the defining properties of an electron, and there is an incredibly important one that we did not mention at all. Electrons have a property known as *electric charge*, which you may well have heard of. If you have not heard of it, do not worry, as for our present purposes, we can simply assume that nature has dictated that there be such a property as electric charge, which can be carried by particles. This charge comes in two types, which are conventionally referred to as *positive* and *negative*, but in fact any other labels would suffice. It then turns out that all fundamental particles in nature can carry electric charge, which may be positive or negative in general. Furthermore, different particles can carry different *amounts* of charge, in a similar way to how particles can have different masses. If a particle has no charge at all, we say that its charge is *zero* or that the particle is *electrically neutral*.

To fully describe a particle, we must give its mass, as well as the electric charge that it carries. I am not getting too technical in explaining the units we use for measuring charge nor how we can measure it directly. But it turns out that the electric charge on the electron is negative, as is the case for the muon and tauon. The amount of charge carried by these particles is identical. The quarks inside the proton and neutron also carry electric charge, and this is so arranged that the neutron ends up having zero charge, whereas the proton carries a positive charge that is exactly opposite to the charge on the electron. This itself has an interesting consequence: before atoms react with other atoms to make molecules or other composite structures, they exist in nature such that the atom is overall electrically neutral. This immediately tells us that the number of electrons in an atom must be exactly the same as the number of protons in the nucleus: the latter carry a positive charge, which will exactly cancel out with the charge of the electrons. The neutrons in the nucleus will then not change this, given that they are also electrically neutral (hence the name).

All of the earlier remarks would be irrelevant if it were not for the following. If we bring different charged particles together, then they exert an influence on each other. Consider the example in Figure 3.2(a), in which a proton is brought close to another proton, such that both of them are stationary. Both of these particles have a positive electric charge, and what happens in practice is that the protons start to move away from each other. Given that they have been at rest and are now moving, this means that their velocity has changed and thus that they have accelerated. The ideas of the previous chapter then tell us that there must be a force that causes this acceleration. Thus, *charged particles create forces*. In this case, we call this a *repulsive force*, given that we can think of each proton as repelling the other. Figure 3.2(b) shows what happens if we bring two electrons together, and we see that there is again a repulsive force.

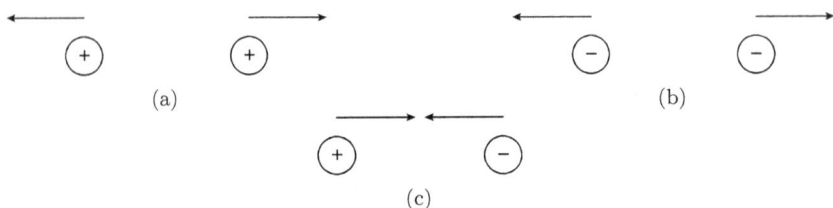

Fig. 3.2. (a) Two positively charged particles (e.g. protons) will repel each other if brought together, where the arrows denote the forces felt by each particle; (b) the same will happen with two negatively charged particles (e.g. electrons); (c) oppositely charged particles will attract each other instead.

However, something different happens if we consider a proton and an electron, as shown in Figure 3.2(c). In that case, such an experiment would show that the particles would move *towards* each other rather than away so that this instead corresponds to an *attractive force*.

We have here considered combining only protons and/or electrons, but the results generalise to all types of fundamental particles, regardless of the amount of charge they carry. Indeed, there is a simple way to summarise Figure 3.2. We can say that any two charged particles exert a force on each other, such that like charges (both positive or both negative) repel each other and opposite charges attract. When talking about personal relationships, the phrase *opposites attract* can be very dubious indeed. But when talking about electric charges, it is certainly true! To complete the picture, we can note for completeness that if we bring together a neutral particle (e.g. a neutron) with a charged particle (e.g. a proton or neutron), then there is no observed force. Thus, *both* particles have to be charged to see a force.

The earlier statements can be made mathematically precise, in that we can write an equation that tells us, if two charged particles are near each other, what force is experienced by each one. Note that the forces in Figure 3.2 can be represented by arrows, similar to the velocities we talked about in the previous chapter. Thus the equations for the force between two electric charges must tell us both the size of the force on each particle and its direction. The equation for the force in each case does this, and it depends (unsurprisingly) on the amount of charge each particle has and also how far apart the charges are. This is perhaps obvious: if we take the charged particles further and further apart, it does not feel right that they should "notice" each other, once they become infinitely separated. Thus, we expect the force to somehow get smaller as the distance between the charged particles increases, and indeed it does.

We now have everything we need to explain why electrons orbit the nucleus in atoms. Recall that the nucleus contains protons, which are positively charged. The electrons are negatively charged, and thus feel an attractive *electric force* towards the nucleus. It is this force that causes the electrons to be attracted to, and ultimately orbit, the nucleus, and we can even write down the relevant mathematics that describes this more precisely. However, even though our knowledge of electric charges is quite practical at this point, we still haven't really

solved the issue of where forces ultimately come from. It is fine to
say that electric charges create electric forces, but what does this
actually entail? Another objection is more philosophical. We stated
that it is possible to write down a formula describing the force on a
charged particle due to another charged particle that is near it and
that this force depends on the distance between the particles. It follows, then, that if we move one of the particles by a small amount,
the force between them changes, *regardless of how far apart the particles are*. Imagine taking this to the extreme: let us put a proton
in Piccadilly Circus and then put another one on the other side of
the universe, billions of galaxies away. The above description suggests that if we wiggle our Piccadilly particle, then the particle in
outer space will suddenly experience a different force to the one it
had before. Can this really be true? I mean really? This idea was certainly highly distasteful to physicists and philosophers in the past,
who sniffily referred to it as *action at a distance*. In more modern
scientific language, we would call such effects *non-local*, where the
terminology is closely related to that used in everyday life. If we
talk about *local* things, we mean things down the road, next door,
or similarly close to us. Likewise, by a *local* theory in physics, we
mean one in which objects can only influence things that are right
next to them, such that there is no sense in which things happening at one place lead to sudden jumps in the behaviour far away.
Our modern understanding of charged particles is indeed local and
removes the need to think about action at a distance. Let us see how
this works.

3.3   Electric Fields

If it is to be true that there are no non-local effects when electric
charges create forces, then there must be something "in between"
the charges that can account for the fact that some sort of influence
travels from one to the other. Our current way of thinking about this
is that a given charged particle creates something called an *electric
field*. The word *field* is used generally in physics to mean a quantity
that is defined at every point in space and at all times. As a simple
example, let us consider the room you are sitting in as you read
this book. If you are not in a room, then let us take some room-sized

region around where you are sitting. We can then consider taking the temperature of the room. Typically, rooms are hotter in some places (e.g. by the radiator) than in others (e.g. next to the window), and thus the temperature will be different at every point in the room at any given time. As time progresses, some parts of the room may get warmer, and some parts may cool down. Thus, in general, the temperature will depend on what time we measure it, as well as the location in space. We would then talk about a "temperature field" throughout the room.

The electric field turns out to be more complicated than the temperature example. Whereas the temperature at each point is a simple number (e.g. 30°C), the electric field has to have a size and a direction associated with it. Otherwise, it cannot describe electric forces, which themselves have a direction as well as a size. We need not worry – there are equations in our modern theory of electric charges that tell us precisely how to calculate the field associated with a charged particle. It does indeed have a size and a direction associated with it so that we can represent the electric field at any point in space (and a given time) as an arrow, whose direction points in the direction of the electric field at that point, and whose size is related to the overall size of the field. In Figure 3.3, I have used a computer program to draw what the electric field of a point charge looks like at different points away from the charge (which is at the centre of the diagram). We see that the field always points outwards from the charge and that the size of the field (as represented by the size of the arrow at each point) decreases as we move away from the charge, in line with the above comments. In Figure 3.3(b), we show the equivalent electric field generated by a negatively charged particle. This looks similar to Figure 3.3(a), except the arrows now point *towards* the charge. The same equations that tell us how the fields for a single charge behave also tell us that we can find the electric field for a more complicated set of charges by simply combining the fields due to individual charges in an appropriate way. I have calculated an example in Figure 3.3(c), which shows the total electric field generated by a pair of oppositely charged particles that are separated by a small distance (in physics language, this is usually called a *dipole*). This is clearly more complicated than the simpler examples of Figures 3.3(a) and 3.3(b). However, the same principles apply, namely that the electric field points outwards from positive

(a)

(b)

(c)

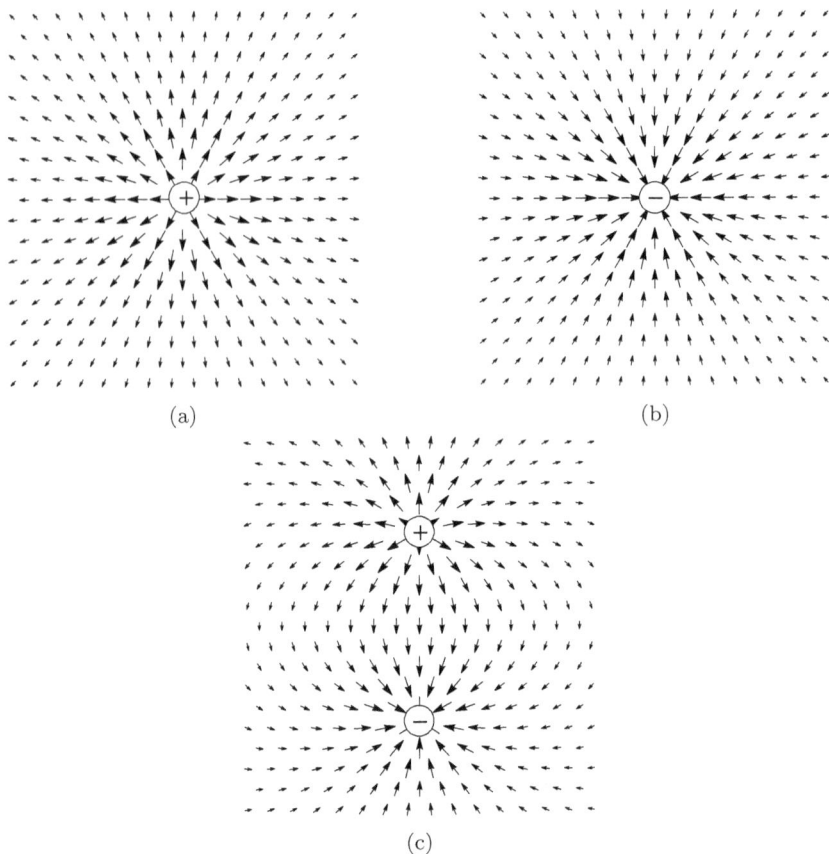

Fig. 3.3. (a) The electric field of a positive point charge, where the direction and size of each arrow gives the direction and size of the field at each point; (b) similar to part (a), but for a negative charge; (c) the electric field of a pair of opposite charges separated by a small distance.

charges and inwards towards negative charges. It also decreases in size as we move further away from either of the charges.

Once we have the notion of an electric field that is created by charges, we can then talk about the electric force, as follows. A given charged particle placed in an electric field feels a force, where there is a precise equation that relates this force to the electric field itself. It turns out that positive charges feel a force along the direction of the field and vice versa for negative charges. Also, a given charged particle does not feel any force due to its *own* electric field: it only feels the forces due to other particles.[1] Thus, in a very well-defined sense, the electric field can be taken as the ultimate origin of electric forces, once we are happy (and we are!) with the idea that charges can create fields. Note that the field concept resolves the dilemma we had above involving crazy non-local behaviour. A given electric charge creates a field filling all space. Moving the charge then disturbs the field, such that this disturbance spreads gradually outwards from the charge, until it reaches other charged particles that may be located elsewhere. At no point is something instantaneously influencing something far away. There is always some sort of local disturbance in the field that propagates outwards such that the field at a given point only ever directly influences the field in a region around that point. A good analogy of this is to think of two corks in a bowl of water. If you push one cork, it will create ripples in the water that gradually spread out until they reach the second cork, causing it to move. It is not true that moving one cork instantaneously causes the other one to move. Rather, the presence of the water in between the corks carries the disturbance from one cork to another, eventually causing the latter to move. The corks and water in this analogy correspond to the charged particles and electric field, respectively.

In this section, we have seen that particles can carry electric charge and that this creates electric fields, which create electric forces on other charged particles. So far, however, our charges have been assumed to be stationary, i.e. not moving. Further strange behaviour occurs when the charges indeed move, as we now discuss.

3.4 Magnetism

I am guessing that you have seen magnetism before, and examples of it in everyday life are not hard to find. It barely seems possible to visit any kind of tourist attraction, for example, without encountering fridge magnets in the gift shop. These contain some sort of metal that sticks to your fridge door (also made of metal), even when you take your hand away. You may also have seen *bar magnets*, such as those shown in Figure 3.4. If so, what you may recall about these objects is

---

[1]One may show very generally in physics that objects which can exert a force on themselves lead to nonsensical results.

Fig. 3.4. A bar magnet has two ends, which are usually called *north* and *south poles*. If two north or two south poles are brought together, they repel, whereas unlike poles attract.

that they have two distinct ends or *poles*, which are typically labelled by the letters N and S, short for north and south, respectively. If you push two like poles together (i.e. two norths or two souths), you can feel a strong resistance, showing that they repel each other. On the other hand, bringing two unlike poles together (a north with a south) creates an attractive force. My toddler has already gotten very used to this idea, as the carriages on his wooden train set contain magnets on the ends so that they can be made to stick together. If, however, he pushes the wrong ends of two carriages together, they will not stick, such that one of the carriages must be turned round. This tells us that each carriage has a north pole at one end and a south pole at the other: if the carriages won't stick, it means two north poles (or two souths) are next to each other. Turning one of the carriages replaces one of the north poles, say, with a south, so that the carriages stick after all.

Bar magnets are electrically neutral overall, in that they are made of atoms that contain equal numbers of positive charges (protons) and negative charges (electrons). Hence, the force between magnets cannot be an electric force but must be something else. It is called a *magnetic force*, and one of the problems with observing it in bar magnets is that the physics of what is going on inside them turns out to be surprisingly complicated (for what it's worth, I explain this as part of a second-year university physics course that I currently teach). This may explain why, for most of our collective history as a species, magnetism was almost completely misunderstood. Given the ideas of the previous section, however, it is actually very easy to state what our modern understanding of magnetism is, as follows.

We saw in the previous section that particles can carry electric charge and that this creates electric fields. If we have a collection of charged particles, these will create an overall electric field, such that each charge experiences a force arising from the fields created by other charges. The situation in magnetism is very similar, provided we supplement the above discussion with a second fact: when charged particles move, they create a *magnetic field*, in addition to the electric field. Like the electric field, the magnetic field potentially fills all of space, and at any given point, it has both a size (the "strength" of the magnetic field) and a direction. We can thus plot the magnetic field of a given collection of moving charged particles using arrows, as we did for the electric field due to stationary charges in the previous section. In Figure 3.5, I have plotted the magnetic field due to a bar magnet. The physics going on inside such a material is really rather complicated, but it ultimately involves lots of charged particles moving inside the material, in such a way as to create a magnetic field. I have plotted this in Figure 3.5, and the result may look familiar, in that we have just seen something very like it. Comparing with Figure 3.3(c), we see that the *magnetic* field of a bar magnet looks very like the *electric* field of a pair of electric charges. It is for this reason that we can think of the north and south poles of a bar magnet as being a bit like magnetic versions of electric charges. There is a crucial difference, however, in that cutting the bar magnet in two always creates new north and south poles, such that it is not actually possible to separate isolated magnetic charges. We return to this point in Chapter 10.

In the case of the electric field, the direction of the field arrows told us the direction that a positive charged particle would move, if it were placed in the field. Another way of saying this is that the force that the electric field creates is always in the same direction as the field itself. This is not true for magnetic fields, such that the interpretation of the field arrows becomes a bit more cumbersome. We can write down an equation for the magnetic force, and it implies that stationary charged particles do not feel any magnetic

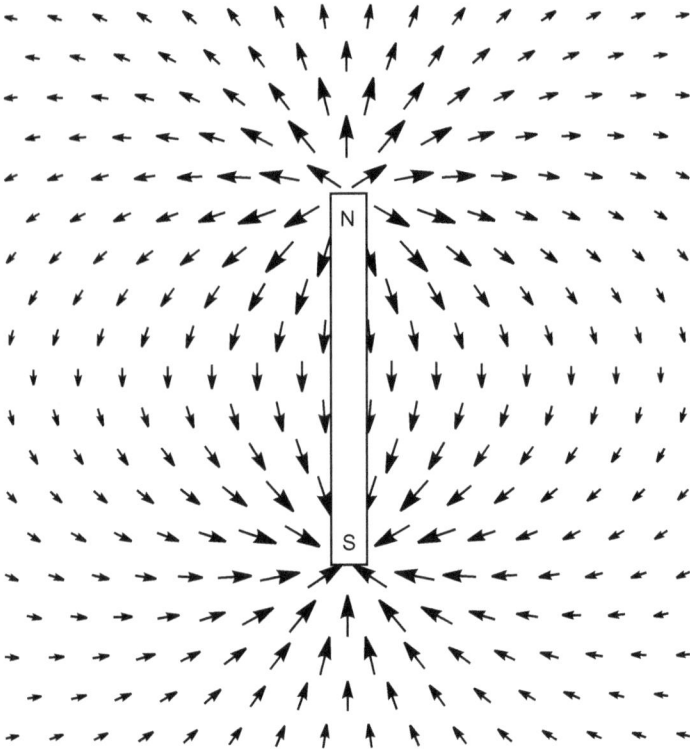

Fig. 3.5.   The magnetic field due to a bar magnet, where the north (N) and south (S) poles are labelled.

force at all, even if there is a magnetic field. Moving charged particles do indeed feel a force, such that this is at right angles to the direction that the particle is moving. Put more simply, magnetic fields tend to make charged particles want to move in circles. Together with electric fields, they can then be used to make charged particles move in any direction we want, and this is used to good effect in particle accelerator experiments such as the Large Hadron Collider at CERN, in which charged particles (protons) are made to move around a ring.

As for electric fields, we can write equations that describe how magnetic fields are generated by moving charged particles, where the latter are also more commonly known as *currents*. However, what may or may not be obvious to you is that it is not possible to write down a theory of electricity alone nor of magnetism. Rather, electricity and magnetism must be combined into a *single theory*. After all, the ultimate origin of both electric and magnetic fields is that charged particles exist in nature. The most general thing these can be doing is moving about, in which case it is automatically true that both electric and magnetic fields will be generated. The combined theory is known as *electromagnetism* and took thousands of years to get right, before finally settling down to something like our modern understanding in the late 1800s. Arguably the main reason for this long history was the need for very detailed experiments in order to establish the correct equations, coupled with the fact that some of the most commonly occurring magnetic materials in nature – such as the magnetised iron used in bar magnets – are actually very complicated from a pure physics point of view. Thus, the dual origin of electricity and magnetism was obscured by circumstances.

3.5 Maxwell's Equations and Light Waves

Although we can choose to think of separate electric and magnetic forces, the fact that these must be combined means that we usually think of a single *electromagnetic force*, which encompasses both effects. There are then two types of equation that are needed to define the complete theory of electromagnetism. First, we must write down the equations that say how both electric and magnetic fields are generated by (moving) charges. The full set of equations of this type is known as *Maxwell's equations*, after the physicist who discovered the final piece of the puzzle joining electricity and magnetism together. A point I often make to my own students, however, is that this puzzle was completed over thousands of years and over many different continents. Knowledge from virtually every culture on Earth gradually mingled and was distilled into the final theory that Maxwell wrote. Since then, both physicists and mathematicians have found new ways to write Maxwell's equations, each of which has taught us something new about the theory.

In addition to Maxwell's equations, we must also say what the force is on a given (moving) charged particle, if it experiences a given electric and/or magnetic field. These and the above equations are typically encountered for the first time in the first or second year of a university physics degree, and in the traditional language, there are five short equations, if expressed in a particularly dense mathematical language. However, the sheer range of behaviour – and *types* of behaviours – that is predicted as a consequence of these equations is breathtaking. Virtually everything we directly experience in our lives can be traced to electromagnetism, not least due to the fact that the forces involved in our five senses are all electromagnetic in origin. Furthermore, our perception of the world around us crucially depends on the thoughts we construct in our minds, which can be directly traced to electric signals in our brains. Even our emotions are electromagnetic: all chemical reactions involve charged particles (e.g. electrons) moving from one place to another so that molecules can split and combine to make new molecules, etc. The constant chemical reactions going on in our own bodies certainly include the various hormones that regulate our emotional selves. Going beyond this, the chemical reactions that keep us alive from moment to moment are also electromagnetic in origin! If we then turn to other species than humans, we see that it is not too much of an exaggeration to state that all of chemistry and biology are essentially consequences of electromagnetism.

To get more of a feeling for quite how widespread electromagnetism is, imagine what you might do in a typical day. You may be woken up by an electric alarm clock or by the alarm on some other electrical device (such as a phone or radio). Upon waking, you may open the curtains and notice that the sky is blue, contrasting this with the red sky that you observed the evening before. If, as more often happens in the UK, the sky is cloudy, it will look white and/or grey depending on how heavy the cloud is. As you get ready to leave the house, you will probably not even notice that you are able to pick things and hold them nor that you do not fall through the floor. You may use various appliances to wash, groom, and/or feed yourself, family, or pets, before leaving the house. Upon going anywhere (e.g. work or school), you may take public transport or sit in a car. Some of this transport will be powered directly by electricity. Other transport may rely on burning fuel (a chemical reaction) but in a highly controlled way such that the energy extracted can be used to drive an electric motor. Modern trains in big cities may allow you to pay by tapping your bank card on a reader, and you may also use such contactless payment if you buy a coffee or snack throughout the day. Your school or work environment may feature computers heavily. Even if it doesn't, you are likely to communicate throughout the day using a smartphone. Amazingly, you can get information from anywhere else on Earth this way, due to a global network of devices transmitting/receiving information, which themselves are linked with satellites in space. After work or school, you may do some exercise or play sport, read books, play computer games, cook food with an oven, watch television, listen to the radio, stroke the fluffy hair of a pet or soft toy, talk to other people, tend a garden, or read a book. Before sleep, you may notice that the sky has gone dark before turning on the lights, enjoy a hot drink, think thoughts about your day, experience a feeling of sleepiness due to hormones in your body, and feel the comforting feeling of a blanket or duvet. While asleep, you may experience dreams, in which increasingly surreal images flicker before your mind's eye, sparking strong emotional responses. We can perhaps stop here, but the heavy-handed point I am trying to make is that all of the above things involve electromagnetism: *every single one of them!* When learning physics the "proper" way, complete with pages of equations and algebra, it is highly common for students to lose sight of the big picture and thus to not quite realize how amazing the subject they are studying is. It is utterly remarkable that the incredibly rich and diverse experience that we call life is not accidental but a consequence of very few fundamental principles or equations. That so much can come from apparently so little is the awesome power of modern physics, where – in my highly subjective opinion – we can date what we call "modern physics" from the first appearance of Maxwell's equations in the later 1800s.

Historically, one of the most immediate uses of Maxwell's equations was to explain the nature of light. You will be familiar with light and its absence from your everyday life. It is the "thing" that bounces off objects and enters our eyes so that we can "see" the objects. You may well have created light by shining a torch, turning on the lights in your home, or lighting a candle. There is also a sense in which light can travel. If you shine a torch up into the sky, you are sending a "beam" of light away from you. Likewise, we are able to see stars in the night sky because the light they emit has travelled to reach us. But what is light actually made of? It turns out that Maxwell's equations give us the answer. One of the things we can do with them is to write the form the equations take when there are no charged particles in some region of space. Then, we can ask what the possible solutions of these equations look like. By a solution, we mean a particular way of arranging the electric and magnetic fields in space, plus how they vary as time passes, and where the equations tell us that this could actually happen. It turns out that there are (infinitely!) many possible solutions, and I have drawn one particular example in Figure 3.6. In part (a) of the figure, we show the electric field of our particular solution in blue and the magnetic field in green. We can see that the magnetic field is at right angles to the electric field. There is also a particular direction (towards the right-hand side), such that both the electric and magnetic fields are always at right angles to it. Finally, we see that the size of the field is not constant as we go along this direction but *oscillates*, getting larger and smaller in a regular pattern. There is more than just this going on, though. Figure 3.6(a) shows the electric and magnetic fields at a particular time. We can then let time pass and ask what happens. An example is shown in Figure 3.6(b), which reveals that, at a later time, the pattern of Figure 3.6(a) has moved forwards along the direction that is at right angles to the fields. Put another way, our solution is travelling along the direction that goes towards the right in the figure. What's more, the shape of the pattern – that it oscillates in an undulating way as we go along the direction of travel – means it is a particular example of something called a *wave*. You will almost certainly have seen many other types of waves, which occur all over physics. Arguably the most familiar example is that of waves in the sea, which we can make mathematically precise by writing equations that say how the height of the sea varies at different places. This equation will have solutions which correspond to similar undulating patterns to those seen in Figure 3.6. The case of an electromagnetic wave is more complicated, however: whereas a water wave has a single thing that is oscillating (the height of the water), an electromagnetic wave has *two things* – the electric and magnetic fields – that are oscillating together at the same time. This turns out to be a direct consequence of Maxwell's equations, which stipulate that changing electric fields cause changing magnetic fields, and vice versa. One way to view the progress of an electromagnetic wave is then that a particular type of varying electric field creates a varying magnetic field that in turn creates a varying electric field, and so on.

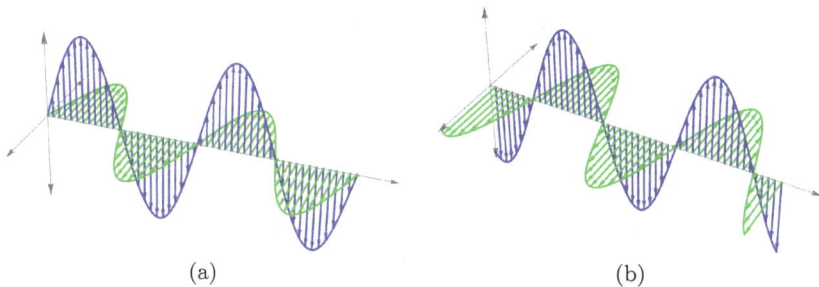

(a)   (b)

Fig. 3.6. (a) An electromagnetic wave, as predicted by Maxwell's equations. The electric and magnetic fields are shown in blue and green, respectively. Part (b) shows the same wave solution, but where some time has passed with respect to part (a).

Figure 3.6 shows a particular electromagnetic wave, but there are many other possible waves. First, the pattern repeats after a certain distance along the direction of travel, which is known in physics parlance as the *wavelength*. Maxwell's equations tell us that waves of any wavelength are valid solutions. Second, we have considered a wave moving in a particular direction, but in fact similar wave solutions moving in any direction we choose are possible. Third, there may be waves which are not associated with a particular fixed direction of travel but instead spread out in all directions. You will be familiar with this in the case of water waves: if you drop a pebble in a pond, you will see circular waves spreading out, and my claim is that similar wave "shapes" exist for electromagnetic waves.

When the wave solutions of Maxwell's equations were found, it was quickly realised that they had the right properties to correspond to light. One reason for this is that it was already known that light travels at a finite speed. The equations, in showing that wave-like configurations of electric and magnetic fields exist, must also tell us the speed of the wave. The predicted value (itself given in terms of certain fundamental properties of electric and magnetic fields) agrees with the measured speed of light. It was also realised that the different "colours" of light that we see (e.g. in a rainbow) correspond to electromagnetic waves of different wavelengths. Indeed, in making this identification, a lot of other things were also explained. It was known that there were many types of waves like light, such that the wavelength was either larger or smaller. In case you are interested, the wavelength of visible light ranges from 0.0000004 m to about 0.0000007 m, where red light is at the upper end of this range and blue light at the lower end. Waves with progressively longer wavelengths are called *infrared, microwaves* (whose wavelength is a few centimetre), and finally *radio waves*. Waves whose wavelength gets smaller than visible light are known as *ultraviolet, X-rays*, and finally *gamma rays*. You may already be familiar with some of these names, and the fact that the names are all different is a consequence of the fact that these types of waves were discovered at different times historically, before it was known that they were different manifestations of a single type of thing: an electromagnetic wave. The full set of such things is nowadays known as the *electromagnetic spectrum*. Given that the only thing separating each type of wave is the wavelength, it is clearly a matter of arbitrary convention where the boundaries between different types of waves are.

3.6   The Speed of Light

Earlier I very casually mentioned that Maxwell's equations, in showing that light is an electromagnetic wave, also predict the speed of light. This turns out to be very fast indeed – around 300000000 m/s – but there is also something weird about this: the speed of light turns out to be the same, *no matter how fast you are travelling*. To see why this is so strange, let us recall that in the previous chapter, I made a point of saying that Newtonian mechanics tells us that if I am travelling at some speed towards something that is moving back at me with a different speed, I would see that object as travelling at an effective speed given by simply adding our speeds together. This *relative velocity* property is familiar to some (not all!) people from everyday life but can ultimately be derived from the equations of Newtonian mechanics.

Imagine that I shine a torch towards you, and assume that you have a device that can measure the speed of light as the torch beam hits you. If you are not moving at all, you would register a light speed of 300000000 m/s, namely the light speed given earlier. Now imagine that you are able to move towards the torch at the same speed at which the light is moving towards you. Newtonian mechanics tells us that your device would measure a light speed of *twice* the one above, i.e. 600000000 m/s. If you were to perform this impossible experiment, however, your device would measure the same light speed of 300000000 m/s. Indeed, the same result would be obtained, no matter how fast you are travelling, a fact which is completely at odds with the Newtonian description of nature. This caused great confusion when it first appeared, such that very detailed and ingenious experiments were performed, aimed at finding differences in the speed of light under different circumstances. However, all these experiments did was verify to very high precision that the speed of light is indeed always the same.[2] Thus, Newtonian mechanics *cannot be a correct description of nature.*

At this point, whether or not this conclusion bothers you is entirely up to you. Indeed, I expect you to have no vested interest in whether a particular set of equations happens to describe our world or a different one. After all, surely it must be possible to patch up Newton's laws of motion so that they can take account of the fixed speed of light. This is exactly what we describe in the following chapter, but it is far more than a simple tweak. The consequences of reconciling electromagnetism with the laws of motion of objects are so far-reaching that they profoundly affect what we even mean by space and time!

Summary

We may summarise this chapter as follows:

- Things around us are made of atoms, which are made of smaller particles, such as electrons, protons, and neutrons.
- All of these – plus related particles – can be reduced to a set of *fundamental matter particles*, which act as building blocks for all the rest.
- Fundamental particles can carry *electric charge*.

---

[2]We should be very careful here and point out that we are talking about the speed of light in a vacuum. Inside materials, the speed of light can effectively change, but this is due to the light interacting with particles in the material, rather than some more fundamental change in the nature of light.

- Charged particles create electric and magnetic fields, which in turn are responsible for producing electric and magnetic forces on other charged particles.
- The equations describing these fields (Maxwell's equations) predict the existence of *electromagnetic waves*, namely oscillating patterns of electric and magnetic fields that can travel through space and which include visible light.
- The speed of light is predicted to be the same no matter how fast you are moving. This contradicts Newtonian mechanics.

Chapter 4

Nothing Is Absolute

In Chapter 2, we looked at how Newtonian mechanics aims to describe how objects made of matter behave, when they are subjected to forces. However, we have now looked at the theory describing a single type of force – electromagnetism – and found that this immediately leads to a contradiction with the consequences of Newton's laws. Historically, this would have come as a massive shock to people, not least due to the fact that Newton's laws had themselves correctly predicted the results of a large number of experiments over hundreds of years. They also describe – to what we now know is merely a very good approximation – the motion of objects that are not on our planet, such as other planets, stars and even whole (clusters of) galaxies. This explains why, towards the end of the nineteenth century, vast amounts of money and effort were expended in trying to *disprove* Maxwell's equations or at least the fact that the speed of light should be constant no matter how fast you are travelling. Some of the experiments used for this purpose were so ingenious that they are still in use today for different reasons. They did not, however, succeed in disproving Maxwell's theory. Instead, an alternative theory of how objects move was proposed that is fully consistent with Maxwell's equations. To introduce it, we need to look in more detail about a curious feature of Newton's theory that we did not yet draw attention to.

4.1   The Principle of Relativity

As we saw in Chapter 3, Newton's laws describe how objects move if there are forces acting on them. All objects have a *velocity* that describes how fast they are moving and in which direction. If the velocity is zero, then the object is not moving so that saying an object has a velocity includes both the stationary and moving cases. Objects also have an *acceleration*, which tells us how the velocity is changing. If the acceleration is zero, this does not mean that a given object is not moving at all. Rather, it means that its velocity is not changing so that it coasts in a particular direction at a constant speed, which may or may not be zero! We also saw that Newton's laws tell us that the acceleration is directly related to the force acting on an object, and this was the key equation that tells us "how objects move".

Now let us consider a situation that you may have encountered before in your everyday life. Imagine that you are sat in a train carriage, that is travelling at a constant speed, and that you look across at another train carriage attached to a different train, that appears to be moving from where you are sitting. How can you be genuinely sure that it is not yourself that is moving, rather than the other train? A variant of this situation is that you look across at the other train and see that it is not moving. But how can you be sure in that case that you are not actually *both* moving, such that the other train is moving at the same speed (and in the same direction) as yourself?

In our everyday lives, we can resolve such issues by looking at our wider field of vision, rather than just at the other train. You would see a background of trees or buildings, and you would be able to see that the other train is indeed moving past them. So imagine a more radical situation: let us replace the trains by spaceships on a blank background (i.e. so that you cannot see any stars). If you see another spaceship, you would have no way of knowing whether one – or both of you – was moving, at least if you are moving at constant velocity. All that you would be able to observe is the motion of the other spaceship *relative to yourself*. It is entirely possible that you might *both* be moving, but with equal velocities such that there is no relative motion. You would then see the other spaceship as apparently stationary.

In Newton's theory, the above statements can be made very precise. One may in fact prove mathematically – using similar mathematics to that described in Chapter 2 when talking about how to add together velocities – that the acceleration of an object is unchanged if you move to a train carriage (or similar moving object) that is moving with constant velocity. To visualise an example of this, imagine we throw a ball up in the air while waiting at a station for our train to arrive. As you well know, this ball will slow down and stop, before returning to your hand (or the floor). The physical reason for this is that the ball experiences the force of gravity. We know that forces cause accelerations, and it is this acceleration that slows down the ball so that it stops, before moving downwards with ever-increasing speed. Now imagine that you get on the train but give your ball to someone else who remains on the station platform. If they throw it upwards as your train moves away, the ball will slow down, stop, and move downwards again in exactly the same way as it did before. The only effect of your motion is that you will see the other person and ball move away from you or, alternatively, yourself move away from them. But this doesn't affect what the ball does in the up and down directions. In other words, the force of gravity and hence the acceleration of the ball *remain the same*, no matter how fast you are moving[1]!

Given that forces and accelerations look the same no matter what (constant) velocity we might be travelling at, a simpler yet grander way of stating the above ideas is as follows. First, we tend to use the fancy word *observer* to talk about the people in the above analogy. That is, an observer is a person equipped with all kinds of hypothetical devices that can precisely measure what given objects are doing. Then we can summarise our discussion by saying that *the laws of physics look the same, for all observers that move with a constant velocity.*

This is a very abstract idea but has become so important in modern physics that it has a fancy name: the *principle of relativity*. It is not, however, a particularly new idea. In Western culture, it was first described in 1632 by Galileo, and thus the principle of relativity in Newtonian mechanics is widely known as *Galilean relativity.*

---

[1]We have to be careful in this analogy and stress again that we are assuming that the train moves at constant speed, which is not generally true for trains as they pull away from stations. But it is only an analogy after all.

It is this principle that forbids us from knowing whether or not we are moving with a given constant velocity. In order to find this out, we would have to do some sort of experiment, and then use the laws of physics to interpret the results of this experiment, and thus to tell us how fast we are moving, and in which direction. But the laws of physics do not care about this, and thus it is *literally impossible* to discern whether or not we are moving at a constant velocity. Put even more bluntly, the question of whether or not we are moving – if we cannot see anything else – is *meaningless*. The only idea of motion that makes any kind of proper sense is that of motion relative to something else. If we see a train carriage that is moving at constant velocity, and we know from experiments in our own carriage that we are not accelerating, then we can only ever talk about our velocity relative to the other carriage.

You may or may not care about whether you can tell you are moving or not. It is after all not exactly our highest priority in life. However, there is something very appealing about the very ethos of the principle of relativity, namely that the laws of physics should look the same no matter where we are in the universe or how fast we are travelling. For one thing, it is suggestive of a huge amount of simplification in our attempts to understand the universe: once we know what the laws of physics look like for us personally, we can extrapolate this knowledge to describe everything everywhere. If we trust the principle of relativity, it is also very powerful in constraining the possible laws of physics we can have. We have seen that Newton's laws cannot be correct. In trying to replace them, we have infinitely many equations we could possibly write down, which would have to be tested against experiments to see whether or not they are correct. However, most of the equations we could write down would violate the principle of relativity. Simply demanding that this principle be respected by the laws of physics leads us extremely efficiently to the right laws of motion, and the power of the principle of relativity is still used today in formulating physical theories.

4.2 Einstein's Theory of Special Relativity

In line with the above comments, let us now see how to patch up Newton's laws, an exercise first done (in its full entirety) by Einstein in the early 1900s. There are two features that we want our laws of motion to have:

(1) The laws of physics are the same for all observers with constant velocity (principle of relativity).
(2) The speed of light is the same for all such observers.

Remarkably, this is enough to tell us precisely what the right laws of motion have to be, without us having to add anything else. It is also a particularly abstract way of doing physics. Up to now, we have largely talked about how the laws of motion can be made familiar by appealing to our everyday lives. We are now insisting, by contrast, that we should assume that the above principles are true and then derive the correct laws of motion from them, *regardless of the consequences*. My reason for stressing this point is that our everyday lives are responsible for an enormous amount of prejudice in our minds. We are not to be necessarily blamed for this: our minds are bombarded with huge amounts of information every second, let alone in a typical day, week, month, or year. Unless our brains applied some sort of filter to this information, we would be quite unable to live. Instead, we impose patterns on the world around us that get increasingly complex as our lives go on. We instantly recognise objects such as trees, cars, or cats by their shape; we recognise different people by the sounds their voices make; we expect the length of a room to be the same no matter how fast we walk along it; we expect our experiences of the passage of time to be shared by our friends. So effective are our brains at imposing order in our surroundings that we are seldom aware of the trillions of assumptions that underpin our conscious minds and hence our personalities. Even if we are, there is a very strong tendency for us to believe that certain assumptions must somehow be *true*. "Of course the world works like that", we might say, "because it's *common sense*".

As a budding scientist from an early age, I have often been accused of having no common sense, but it is only in later years that I realised this was something to be proud of. By "common sense" is almost always meant some sort of collective and accepted knowledge, whose underlying assumptions are beyond question. However, the consequences of the two principles above – and hence our modern laws of physics – do not satisfy any reasonable definition of common sense! I will describe situations that are so alien to our everyday experience that it will be very difficult for a novice to accept them. The history of physics is by now so full of such cases that we know the correct way to proceed: every time we encounter a consequence of our physical laws that is at odds with our daily experience, it is our experience that is "wrong" and not the physical laws. More strongly, our everyday experience and "intuition" are incredibly unreliable when it comes to understanding the laws of modern physics. Whether or not a given physical theory is correct must be decided by increasingly complicated experiments and not by what we think or hope is right or wrong.

With this in mind, let us now examine the consequences of the above two principles. The theory that emerges when we do so has become known as the theory of *Special Relativity*, and this name distinguishes it from a more general theory that we see later on. For our present purposes, we can view Special Relativity as a new set of laws for how objects move, which formally replace Newton's laws. Consequently, the new laws of motion change how we must think about space and time, in ways that are much more radical than you might think. We explore what happens to time in the following section.

4.3   Time Dilation

When you are doing something very interesting and/or pleasurable, time seems to pass far too quickly. It feels like we have to remind ourselves to mentally slow down and take things in so that we can appreciate them later as pleasurable memories. This is often called *living in the moment* and is also a key part of mindfulness techniques that have been shown to have a positive impact on our mental health. Conversely, when the circumstances around us are physically or emotionally taxing, or even just plain boring, we are likely to feel as if time is ticking by painfully slowly. We may find ourselves watching the clock, willing it to tick faster, and frustrated that we have no influence on its progress.

While the above comments may suggest that the notion of time is somehow flexible, it is important to note that they relate only to our human perception of time which, given our lack of understanding of key aspects of the brain, is just as unreliable when it comes to how the real world works as any other of our highly fallible senses. No sober person in everyday life would ordinarily suggest that time is somehow *different* for different people. If you buy a watch in Kathmandu, you expect its ticks and tocks to be precisely as far apart in time as if you bought an identical watch in New York. If we saw clocks ticking at different rates, we would simply say that the battery is going flat or that a mechanical timepiece has not been wound properly, which indeed would be the correct explanation for such cases.

As we saw in Chapter 2, the idea that time is somehow "the same" for all observers is known as *absolute time*. It is one of the great prejudices that our minds impose on the world around us, in order to comprehend our existence. For many of us, the notion of absolute time is so self-evident that we would never have any reason at all to question its validity. Nor is it easy to understand what it even means to say that time is not absolute and what the knock-on effects of this statement might be. Some people – whether they realise it or not – have a very deep emotional attachment to the notion of absolute time. If we think about the most upsetting aspects of our own lives, they nearly all involve time as a key factor: the "direction" of time implies ageing, death, and decay, both of ourselves and those we love. We are stuck with memories of painful events in the past, with no ability to change them, and a tendency to endlessly ruminate over what we might have done differently. Even our happy memories are of isolated pleasurable experiences that we can never repeat or revisit, much as we would like to. We experience uncertainty, stress, and fear about what the future holds in store for us, and it takes great skill and practice to cope with this uncertainty, if we find ourselves out of our personal comfort zones.

My ultimate aim here is not to depress you but merely to point out that time and our perception of it is a crucial part of what makes us human. This in turn creates strong feelings about what we feel we are allowed to say about time and also strong reactions against theories that predict that our everyday experience of time is not in fact correct. Special Relativity is one such theory, and it tells us in no uncertain terms that time cannot be absolute. That is, time must tick at different rates for different observers, depending on how fast they are moving. Let us now see how this follows from the two principles stated earlier, using a classic argument involving torches, mirrors, and trains.

Imagine that you have a torch, and you shine it at a mirror as shown in Figure 4.1(a). When you turn the torch on, it creates

Mirror

Torch

(a)

(i) (ii)

(b)

(i)   (ii)

(c)

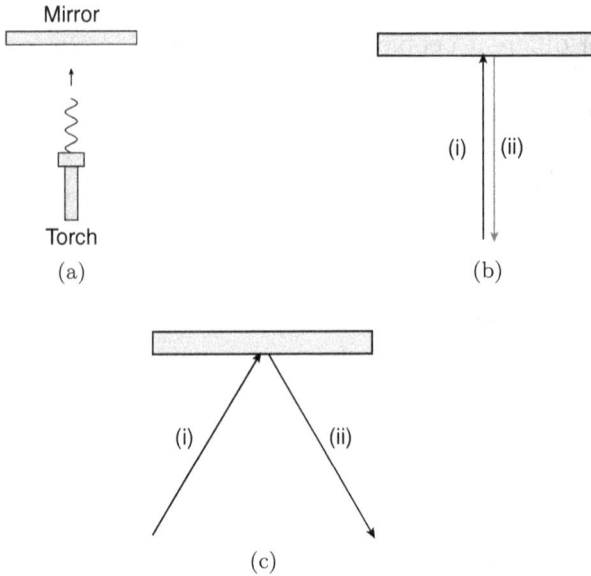

Fig. 4.1.   (a) A torch shining at a mirror creates a beam of light, which travels up the mirror and is then reflected; (b) the paths taken by the two light beams (i) travelling to the mirror and (ii) reflected by the mirror; (c) the same experiment as seen by an observer at a level crossing, who sees the apparatus go past on a train.

a beam of light that travels at a finite speed (predicted by Maxwell's equations) until it hits the mirror. The latter then *reflects* the light, creating a second beam of light that travels back to where the torch is. We will not worry about how this reflection works, which will not be necessary for our purposes. My hope, however, is that you are familiar with the fact that mirrors reflect light from your everyday life. If we want to, we can draw the paths taken by the light that goes as follows: (i) from the torch to the mirror and (ii) from the mirror back to the torch. These are shown as arrows in Figure 4.1(b), which in turn make clear that the light must travel a certain distance in going to the mirror.

So far so good. But let us now complicate matters, by saying that rather than doing this experiment yourself, you have given your torch and mirror to a friend, who stands in a train carriage. You have arranged to meet the train at a level crossing so that the person shines the torch at the mirror as the train passes from left to right in front of your eyes. From your point of view, the beam of light will now travel up to the mirror and back as before. But, because of the extra motion of the train, there will be a sideways component of the light beams: the whole experimental apparatus is moving from left to right, and thus the light beams will also be doing this as the train moves past. For the person on the train, the light beams will appear as in Figure 4.1(b). For yourself at the level crossing, they will instead look like Figure 4.1(c). Nothing so far is particularly unusual – we have simply stated that things look different if you happen to be moving or not. Nor does this contradict the principle of relativity, which says that the laws of physics have to be the same for all observers but still allows for the fact that individual observers may see different things.

Unusual things do happen, however, if we take into account the second principle earlier, namely that the speed of light is the same for all observers. As discussed in Chapter 2, speed measures that you have gone a certain distance in a certain time. The person on the train, and yourself at the level crossing, must measure the same speed of light, according to the rules we laid down for ourselves. But we can clearly see from Figures 4.1(b) and 4.1(c) that the distance travelled by the light according to our two observers is very different. That is, yourself at the level crossing will see that the light has travelled a *larger* distance (the combined lengths of the arrows in Figure 4.1(c)) than the person on the train measures (the combined lengths of the arrows in Figure 4.1(b)). Given that the distance is different, it has to be the case that the period of time measured for the light to get to the mirror and back is different as measured on the train, to at the level crossing. Otherwise, our two observers would not observe the same speed of light!

This argument was first presented to me on a school trip to CERN. I would have been 16 or 17 at the time, and my very enterprising physics teacher had somehow managed to get a university professor to take us out to CERN for a week, which then counted as the "work experience" we were obliged to carry out. Not only this, but the professor also arranged for us to meet several characters at CERN who you would never encounter on any kind of public tour. One of them was the Head of Radiation Safety, whose job was to ensure that all staff – and residents of nearby villages – remained safe at all times, despite the fact that a major particle accelerator was colliding subatomic particles beneath their feet. It was a formidable job and responsibility, requiring a considerable sense of humour. One of his party tricks was to gesture towards a shed and exclaim, "There's the most radioactive building on site!". As we gasped in horror, he would further explain, "It's where the local deer farmers keep their fertiliser[2]". Other highlights included tips on how to halve one's nightly radiation dose if married (sleep alone) and also how to estimate the effect of lost radiation beams on river trout. The point he was repeatedly making is that all life involves risk and that all radiation risks must be placed in a wider context of naturally occurring radioactivity. More widely, he was stressing that physics itself is all around us, if we only know where to look. At some point in this discussion, he saw a look of puzzlement on our faces due to never having heard of time being different for different observers. "But it's easy to see why!", he shouted, before drawing something very like Figure 4.1 on his blackboard. He then went further than this, by explicitly deriving the mathematics of how time has to change, to comply with the twin requirements of relativity and a constant speed of light. It was one of the key formative experiments of my scientific life and has influenced how I think about the subject ever since.

While it is clearly beyond the scope of this book to show you the mathematics that so enthralled me in a dusty office thirty years ago, we can at least be more precise in stating how time has to change for the observers in Figure 4.1. For the person on the train, the light travels a shorter distance. Thus, they must measure a shorter time in order to obtain the same speed of light, than the person at the level crossing. On the other hand, yourself at the level crossing sees the same process as your friend on the train (light bouncing off a mirror), but you will find that it must take longer, in order for the light speed to be the same. If you were to see your friend's clock (or other time-measuring device) as they whizz past on the train, you would thus see their clock as ticking *more slowly* than your own clock. This effect is known as *time dilation*, given that the moving object's time is somehow lengthened according to the observer at the level crossing. But let us be very clear about this: the fact you see their clock as

---

[2]Fertiliser does indeed contain very small amounts of radioactive elements, such as potassium and uranium.

running slowly has nothing to do with batteries going flat or clocks not being wound up properly. *The nature of time has changed*, and it has changed in just such a way as to guarantee that light travels at the same speed for all observers. This is a profound difference with respect to Newton's theory, and one that is very much at odds with our everyday experience. The reason why this experience fools us is that objects have to be moving at speeds very close to the speed of light before we see the effects of time dilation.

I went to some lengths earlier to point out how attached some of us are to our notions of time, given the emotional impact that changes to this worldview can have. Indeed, it seems still to be the case today that some people become actively angry at the notion that time may not be the same for all observers. Similar to any working scientist with a university email address that is easily findable on the internet, I am regularly contacted by people who wish to claim that Special Relativity is false and that some complicated mechanism is needed to restore absolute time, in a way that remains compatible with a constant speed of light. By far the simplest way out, however, is to say that Special Relativity is correct, and indeed detailed experiments confirm this viewpoint. One such "experiment" involves detecting subatomic particles that are produced in the upper atmosphere. A particular one – the *muon* that we briefly encountered in the previous chapter – would decay very quickly to form an electron and another particle, if left by itself. The time for this to happen is 0.000002 s, and without Special Relativity, it would be a mystery why we should be able to detect such particles in the surface of the Earth: if produced in the upper atmosphere, they would all have decayed before they could reach us. Time dilation resolves the puzzle: the muon is moving close to the speed of light relative to us, such that the time it takes to decay is dilated according to our point of view. It is thus able to reach our detectors on the Earth's surface, which ultimately destroy the muon in a different way! Similar effects happen all the time at the Large Hadron Collider at CERN. Particles that would normally have very short lifetimes suffer time dilation so that they can be observed in large detectors which surround the points at which beams of protons collide with each other. In all such cases (including the muons discussed earlier), Special Relativity allows us to calculate what the dilated lifetime of each particle should be, and we get results that very precisely agree with experiment.

If you have trouble comprehending what it means for time to tick at different rates for different observers, do not worry. However, do not make the mistake of thinking that you somehow need to understand complicated mathematics to fully "get" what is going on. In my experience, university physics students have no trouble grasping the mathematics behind time dilation – which can also be understood by a high school student. Rather, it is the very idea of time dilation that is somehow hard to grasp, and this difficulty is almost always traceable to the fact that we *expect* time to behave differently, or are so constrained by our everyday experience, that we have trouble thinking beyond it. Ultimately, there is nothing particularly special going on. As soon as we demand the principle of relativity and the fact that the speed of light is the same for all observers, time dilation is an *unavoidable consequence*. If we did not have the second condition, then Newton's theory – which has absolute time – would result instead. Once we get used to the idea that it is not our experience that matters but the assumptions that go into our theory, much of our conceptual difficulty goes away. We need no longer insist that the consequences of real physical theories are automatically relatable to our everyday experience, and over time we gradually develop a new "intuition", that comes from applying Special Relativity to real physical examples, with data to confirm them. At least, that's what happened in my own career.

Time dilation is not the only surprise lurking in Special Relativity. We must also rethink our naïve notions of past and future. I made a very simple sketch in Figure 2.2, which shows a line marking out different values of time, where a specific point is called "now" or "the present". Anything to the left of this is called "the past", and anything to the right is called "the future". In our own personal experience, this is yet another of those things that appears so self-evident as to be obviously true and/or unquestionable. That it cannot actually be true follows from the fact, once one analyses the theory of Special Relativity, that the speed of light turns out to be the maximum speed at which objects can travel. What we conventionally call "the past" consists of all those events that could have influenced you today. To do this, they must have been able to send this influence over distances, in order to reach you. However, if something was far enough away from you at an earlier time, it would be unable to send you any information, due to the speed of light not being fast enough.

Likewise, at later times from now, there are events that you will not be able to influence, due to only being able to send information at the speed of light and no faster. What we conventionally call "the future" consists of all events that you can indeed influence, and thus there are areas of the universe at both earlier and later times that are somehow "disconnected" from you, being neither part of your past nor your future. Physicists call this the "elsewhere", and this distinction matters a great deal for ensuring that our current theories are consistent so that nothing can influence its own past.

4.4 Length Contraction

In the previous section, we saw that insisting that the speed of light is the same for all observers means that we have to modify the notion of time very significantly. However, given that speed involves both time and distance, it is natural to ask whether we need to modify our understanding of distances too. Indeed this is the case, and there is an accompanying effect to time dilation, known as *length contraction*, which is again derivable from the equations of Special Relativity. The effect states that if we see an object such as the train earlier moving past us from left to right, it will appear to be flattened or *contracted*, in its direction of motion. An example is shown in Figure 4.2, which shows what would happen if a ball were to travel at close to the speed of light. Part (a) of the figure shows a perfectly spherical ball, which is the shape we would see were the ball at rest. Once the ball is moving as shown in Figure 4.2(b), it is contracted to make an egg-like shape, and this flattening gets more extreme the faster it gets. Note that the equations tell us that the flattening effect is

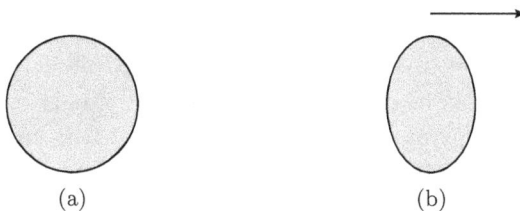

Fig. 4.2. (a) A ball at rest; (b) a ball travelling at close to the speed of light in the direction shown.

only in the direction in which the ball is travelling, not in the other directions. We can see this directly in Figure 4.2(b): the height of the ball remains the same, even when its width reduces.

A simple way to understand why length contraction must exist is to reconsider the muons discussed earlier. When these are produced in the Earth's upper atmosphere, they are travelling very fast. As we saw, time dilation means that their natural lifetime is elongated, such that the muons survive to be detected close to the Earth's surface. However, all of this is what happens according to us. How do things look from the muon's point of view? If we were able to sit on a moving muon, our clock would tick at the same rate as it, and thus we would expect to see it decay on a very short timescale. However, given that we would then see the surface of the Earth as moving towards us (and the muon) at close to the speed of light, the distance from the upper atmosphere to the Earth's surface would effectively be reduced due to length contraction. Hence, the muon does not have as far to go as if it were at rest relative to the Earth, and it thus makes it to the Earth's surface before it can decay! The stories told by an observer who hugs the muon and one on the Earth's surface are very different. But the overall conclusion is very much the same. Physically meaningful things (e.g. whether a muon gets detected on the surface of the Earth or not) must end up having a sensible interpretation, whichever way we look at them. The equations of Special Relativity guarantee for us that this will always be the case.

4.5   Energy and Mass

There is another feature of Special Relativity that is worth drawing attention to, namely a certain relationship between the mass of an object and the energy it has. We have previously seen mass in the context of Newtonian mechanics in Chapter 2 and saw that it measures the inertia of an object. That is, an object with higher mass has more resistance to changes in its motion: if it is stationary, it is harder to get it moving; if it is already moving, it is harder to deflect it. This concept of mass survives in Special Relativity, although the relationship between forces and accelerations becomes more complicated. If we now want to relate mass to the energy of an object, we have to carefully define the latter.

You will be familiar with the concept of energy from everyday life. If a person is said to have more energy, they are less sleepy and more active. Likewise, if an object is moving faster than some other object, we say that it has more energy. This notion can be made very precise in physics, and we say that objects that are moving (have a non-zero velocity) possess *kinetic energy*, whose name derives from the Greek word for motion. In Newtonian mechanics, there is a precise formula we can write down that tells us how much kinetic energy an object has, once we know its mass and its speed. Unsurprisingly, if an object is not moving at all, it has no kinetic energy.

The concept of kinetic energy is a lot more general than it might first appear. For example, sound waves in air, or water waves in the sea, can clearly carry energy. We would say in this case that the air or water has kinetic energy, and this is ultimately traceable to the movement of the individual atoms and molecules. This kinetic energy forms part of the energy of the sound wave. As another example, engineers often talk about *thermal energy*, which is the energy that a given substance has due to being at a given temperature. Hotter things have more energy, and we can make this more precise by carefully defining what we mean by temperature. Our modern understanding is that the temperature of a material is directly related to how fast its constituent atoms or molecules are jiggling about. Thus, thermal energy is also simply kinetic energy in disguise.

As well as kinetic energy, objects can have something called *potential energy*, which is more abstract to think about. One way to understand the need for it is to note that all of our current theories of physics tell us that the total amount of energy in the universe has to remain the same. Physicists call this *conservation of energy*, and it is only one of a number of so-called *conservation laws* that apply to different quantities. Now imagine that I place a cup of coffee precariously close to the edge of my desk, as shown in Figure 4.3(a). The picture shows the typical office of a working theoretical physicist. There are notepads of algebra that may or may not relate to what is written on the whiteboard (by my excellent PhD student Kymani!). And there is a shirt hanging up, in case of a sudden need to be unusually smartly attired. We can also see a phone, which is seldom used nowadays, but which does occasionally loudly ring. This causes a certain amount of surprise in my own person, and I will then attempt to answer the phone, hampered by the usual

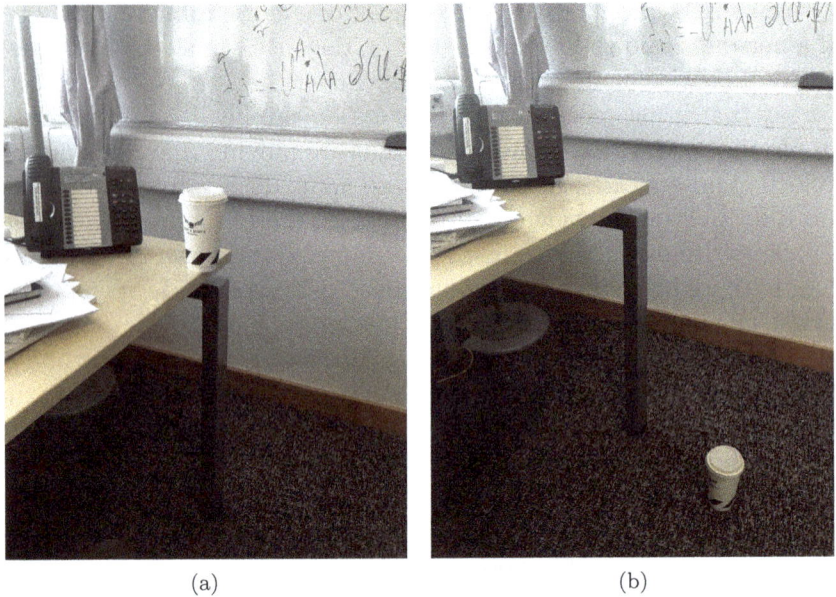

(a)   (b)

Fig. 4.3.   (a) A cup of coffee precariously close to the edge of my desk in East London; (b) a cup of coffee having fallen to the floor.

clumsiness of the theoretical physicist. As my gangly arms reach across the notepads of algebra, I am highly likely to knock the coffee cup from its perch, leading to the situation of Figure 4.3(b) or worse. You will doubtless have knocked many a thing off a table yourself, particularly in your rumbunctious youth. You will therefore know that the cup in Figure 4.3(b) will be *moving* when it hits the floor. From our earlier discussion, this means that it must have kinetic energy. However, if the total amount of energy is to be the same at all times, this energy must have come from somewhere, and indeed it has. We say that the cup in Figure 4.3(a) has a *potential energy*, due to having been raised against a gravitational force. It is this potential energy that gets converted to kinetic energy when the cup falls. Physicists can do a lot with this idea. In the Newtonian theory of gravity, for example, we can write a precise equation for what the potential energy of an object is. By saying that this has to become kinetic energy, we can predict exactly how fast the cup will be moving when it hits the floor! While impressive, this is of course scant consolation for the coffee stain on the carpet.

Going the other way, if we lift something against the force of gravity, we must be creating potential energy, where this energy has come from the effort we expend with our arms, legs, and back. The potential energy of an object experiencing gravity depends upon its mass, which is easy to appreciate. You will know yourself that lifting a piano up a flight of stairs is considerably harder than lifting a marshmallow, even if you had a huge marshmallow that had the same size as a piano. We would say in this case that the piano has more potential energy than the giant marshmallow, once they are both at the top of the stairs.

Although we have talked about gravity here, the idea of potential energy generalises to all other kinds of force. That is, every time we have a force in nature, there is a potential energy associated with moving things against the force. While it is simple to state this, you may find yourself unsatisfied. In the case of kinetic energy, we can see directly that an object is moving, and thus it is obvious to us that it carries some form of energy. For the cup in Figure 4.3(a), we cannot see any obvious form of energy, so where does the potential energy actually reside? To answer this, note that the only force we have examined in any detail so far is that of electromagnetism in the previous chapter. There, we saw that charged particles create electric and magnetic fields. What Maxwell's theory tells us is that these fields can themselves store energy. Thus, this is the origin of electromagnetic potential energy. Likewise, for the gravitational force, massive particles can create a gravitational field, and this is where the potential energy is sitting, before being converted to other forms of energy as needed. We deal with the force of gravity in much more detail in Chapter 7.

Like kinetic energy, potential energy comes in many forms that are often given different names. For example, the sound and water waves we mentioned earlier will carry some potential energy in them, due to the electromagnetic interactions between the constituent parts of the water or air as it gets stretched and compressed. We would typically call this *elastic energy* in that case, and similar words would be used to talk about the energy stored in a stretched rope or spring. Chemists might talk about *chemical energy*, as referring to the potential energy in some system (e.g. a battery) that can be extracted for other purposes. All other forms of energy that you may have heard people talk about (e.g. nuclear energy, heat energy, and mechanical energy) are always some combination of kinetic and potential energies, if thought about in a more fundamental way.

Now that we are more comfortable with different forms of energy, let us see what Special Relativity has to say about them. Regarding potential energy, this idea is more or less the same as it was before: if we have a force, this has a potential energy associated with it that is calculated in a prescribed way. Even if there are no forces, however, objects can have energy. We saw that in Newtonian mechanics the only possibility is kinetic energy and that this must vanish if an object is not moving. However, in Special Relativity, we get a different formula that tells us that objects can still have an energy, even when there are no forces, and if the object is completely stationary. This "extra" form of energy is typically called *rest mass energy* and depends only on the mass of the object and the speed of light. The use of equations in books such as these is typically frowned upon, but I am guessing that you will have heard of Einstein's famous equation

$$E = mc^2,$$

albeit without necessarily knowing what it means. It is none other than the formula that tells us how to calculate the rest mass energy of an object (denoted by E), if we know its mass (m) and the speed of light (c). If you do not understand how to apply this formula, do not worry in the slightest. But it seemed oddly churlish not to mention arguably the most famous equation of the twentieth (or perhaps any) century! We will refer to this as the mass-energy relation, given that it tells us how the mass of an object is related to its rest mass energy.

Simply saying that objects have rest mass energy does not tell us what it actually means. We see some applications of the idea later, but it is perhaps helpful to have given one example already here. In the previous chapter, we saw that atoms contain a mix of protons and neutrons at their core, which we call the *atomic nucleus*. Many experiments have been performed on nuclei, and they can be summarised by saying that when we make a nucleus out of protons and neutrons, the energy of the resulting nucleus is *lower* than that of the separate protons and neutrons, due to the potential energy associated with the forces binding the nucleus together.[3] It follows

---

[3]Unlike kinetic energy, which always has to be positive, potential energy can be positive or negative, where this is associated with whether a force is attractive or repulsive.

from the mass–energy relation that the mass of a nucleus will always be *less* than the mass of the corresponding protons and neutrons, were these to be considered as separate particles. If we were to split the nucleus into separate parts, this energy could be released, and this is what is happening in a nuclear power plant. More generally, one way to think about the mass–energy relation is that it somehow implies that mass and energy are very similar concepts that are to a certain degree interchangeable. We see this idea recurring in what follows.

4.6 Was Newton Wrong and Einstein Right?

In this chapter, we have seen that Newton's laws of motion must be replaced with a completely different set of equations (Special Relativity) that reconcile the principle of relativity with the fact that the speed of light is the same for all observers moving with a constant velocity. It is thus tempting to say that Newton's laws are "wrong" and that Special Relativity is "right", in some sense. What constitutes truth is a question that is probably best left to philosophers, but we might still try and be more precise by saying that Newton's laws do not agree with experiments and that Special Relativity agrees. This is indeed our modern view of what constitutes an effective scientific theory, but it is very important to point out that this does not mean that Newton's laws are useless nor that Special Relativity is ultimately true. To justify these remarks, let us first remind ourselves from earlier that the effects of, e.g. time dilation and length contraction are only noticeable if objects are moving very close to the speed of light. A quick glance around your immediate environment will reveal that most of the objects we see in our everyday lives are not doing this! Thus, Newton's laws are a perfectly valid description of everyday physics, to an extremely good (and quantifiable) approximation. When a sports scientist applies physics to improve an athlete's performance, they use Newton's laws and not Einstein's. Not even Usain Bolt can run close to the speed of light.[4]

This itself leads to an important idea. When we express Special Relativity mathematically as a detailed set of equations, it must

---

[4]In case you are interested, Usain Bolt's top recorded speed is around 0.00000004 times the speed of light. It is a lot more impressive to say that it is almost 30 mph!

somehow be true that, if objects have small speeds, the equations reduce to Newton's laws in a well-defined way. We should see Newton's laws "emerging", if we take the velocity of a given object to zero or, equivalently, if we take the speed of light to be very large. This indeed happens, and it imposes a very powerful constraint on the equations of Special Relativity themselves. In modern parlance, we say that Newton's laws are an *effective description* of Special Relativity at low speeds. However, this also forces us to rethink the notion of whether Einstein's equations are really "true". We used to think that Newton's laws were all there was, until we found that they had to be replaced in what, at the time, would have been very extreme circumstances. How do we know the same is not true of Special Relativity? In other words, might Special Relativity itself be an effective description of a more underlying theory? This is indeed a real possibility, such that – in Western culture at least – the birth of Special Relativity constituted a paradigm shift in how we think about science. We nowadays accept that all of our theories may turn out to be merely effective descriptions of something else, subject to some as yet unperformed experiment revealing that our current theories break down. However, this would have been profoundly shocking to some eighteenth century thinkers who, following the first industrial revolution, would have found it very difficult to accept that our certainty of knowledge about the world around us was suddenly in doubt.

In my experience, the general public is unaware of these aspects of modern science, namely that (i) theories are not only allowed, but expected, to be fallible and (ii) this is not a problem, given that any given set of laws of physics provides an excellent approximation where it applies. An amusing illustration of this a few years ago was the purported discovery of certain particles (neutrinos) travelling faster than the speed of light, which is certainly forbidden in Special Relativity. While making breakfast, I happened to be listening to a flagship BBC radio programme, whose particularly smug presenter revelled in the fact that "Einstein was wrong". The source of his joy – and the implied camaraderie with his listeners – was presumably in the fact that he had never wanted to understand Special Relativity and thus felt vindicated in ignoring it. All of that is fine, of course, unless the smugness happens to be misplaced. No physicist would be surprised to learn that Special Relativity is wrong, merely excited to learn that a more underlying theory has Special Relativity as its effective description! Moreover, no physicist would try to use this more underlying theory in contexts in which Special Relativity suffices. As it turned out, an experimental error meant that the discovery turned out not to be founded after all. But it did at least shine an interesting light on public attitudes to science.

The other key idea we have learnt in this chapter is that trusting our everyday intuition about how the world works is not the right way to do science. We should instead try to find the key principles that have to be true and then use these to derive a mathematical framework that gets them right. In turn, this set of equations may lead to consequences that are so at odds with our feelings of how the world should work that we may have trouble accepting them. Physicists have learnt by now that nature rarely cares about our feelings. When faced with baffling conclusions – such as time ticking at different rates for different people – we should trust the mathematics and let experiments decide whether or not a given theory is true. This is an important message with which to end this chapter. In the following chapter, we see that Newton's laws break down quite independently in a different context, whose consequences will be far weirder than anything we have seen so far.

Summary

A summary of key points from this chapter is as follows:

- The theory of electromagnetism predicts the same speed of light, no matter how fast you are travelling.
- This is inconsistent with Newtonian mechanics, which must therefore be replaced by a more underlying theory.
- This theory is called *Special Relativity* and forces us to rethink our notions of space and time.
- We also learn that particles at rest can have a new type of energy, even when there are no forces acting.
- Newton's laws must emerge from the equations of Special Relativity in a precise way. We say that Newton's laws are an *effective description* of Special Relativity, if objects move much more slowly than the speed of light.

Chapter 5

Little by Little

We arrived at the need for Special Relativity by noting that the constant nature of the speed of light meant Newton's laws needed to be modified. Our discussion at the end of the last chapter tells us that another way to think about this is that Newton's laws break down if objects move very fast, such that their speeds approach that of light. It turns out that this is not the only such circumstance: they must also break down as objects become very small, and the reason for this is that the theory of electromagnetism behaves very badly indeed.

5.1 A Few Small Problems

5.1.1 *Why do atoms exist?*

As we stated in the previous chapter, most of what we see around us is made of atoms. We also gave a simplified picture of what an atom is as follows: it has a central nucleus containing protons and neutrons, around which orbit other particles called electrons. The electrons carry electric charge, and the fact that they are orbiting the nucleus means that they must be accelerating. To see this, recall from Chapter 2 that the acceleration of an object measures how its velocity is changing, where the latter corresponds to speed in a particular direction. Thus, if an object is accelerating, its speed, its direction, or both must be changing. If we consider an object orbiting another one, it is constantly changing its direction. If you don't believe me, imagine I place a chair in the middle of a room and ask you to walk around it (i.e. orbit it). Whether you choose to go clockwise or anti-clockwise, you will certainly have to turn at some point so that you can make it round the chair. In this analogy, you are representing an electron, and the chair is an atomic nucleus. We have thus established that electrons in atoms are indeed accelerating.

Let me now tell you the reason why I am drawing such attention to this aspect of electron motion. There is a consequence of Maxwell's equations that is somewhat tedious to derive but which is nevertheless unavoidable: *if charged particles accelerate, they radiate electromagnetic waves.* We described the existence of electromagnetic waves in Chapter 3. To remind you, they are complicated patterns of electric and magnetic fields – such as that shown in Figure 3.6 – that move at the speed of light. Indeed, light itself is made of electromagnetic waves, as are all other types of radiation on the electromagnetic spectrum, such as X-rays, microwaves, radio waves, and other famous things. However, merely showing that wave solutions of Maxwell's equations exist does not itself imply that such waves will be observed in nature. It may sometimes be the case in physics that certain solutions of the equations exist but that these are not manifested in the particular universe we happen to live in. To fully understand electromagnetic waves, we must say how they can be produced, and this is precisely what the earlier discussion does for us. If we take a collection of charged particles and accelerate them, this will in turn generate electromagnetic waves, which travel away from the accelerated charges. It is certainly the case that the world around us contains accelerating charged particles, and thus we can expect to see electromagnetic waves – and we do! There is a simple everyday application of this. If you have a mobile phone, it will use the battery to create electric fields. These can be used to shake the electrons in a part of the phone known as the *antenna*, and as soon as this happens, electromagnetic waves will be radiated away. By intricately modifying how the electric fields are produced according to how you are speaking, the waves can be made to carry information that can then be received by somebody else, possibly after passing through intermediate devices, such as communications masts and satellites. Indeed, the fact that such complicated communication networks exist around the globe tells us that we very precisely understand how electromagnetic radiation is produced by accelerating charges.

Returning to the case of electrons in an atom, we now know that the electron will radiate away energy as it follows its orbit. If you were following the discussion about potential energy in the previous chapter, you will remember that the total energy of the universe has to remain constant. Thus, if our electron radiates away energy, it must *lower* its potential energy, in order that the total energy remains the same. It turns out that it can do this by orbiting closer to the nucleus and, upon doing so, it will continue to radiate more energy so that it must move closer still. The upshot of this behaviour is that all electrons in all atoms will spiral into the nucleus, as shown in Figure 5.1. Furthermore, the equations tell us that this will happen extremely quickly.

If you have not been following the discussion about (potential) energy, do not worry. The important point is that, according to the equations of electromagnetism, electrons will not actually stay orbiting in atoms for very long. Instead, all atoms are unstable and will quickly collapse! A simple experiment is enough to show us that this conclusion is complete nonsense: all you need to do is notice that you are alive and reading this book. Then the only sensible thing we can conclude is that something is very wrong with the theory of electromagnetism once we get down to the size of an atom.

Of course, this is not the first time that we have seen laws of physics break down. We saw before that Newton's laws of motion had

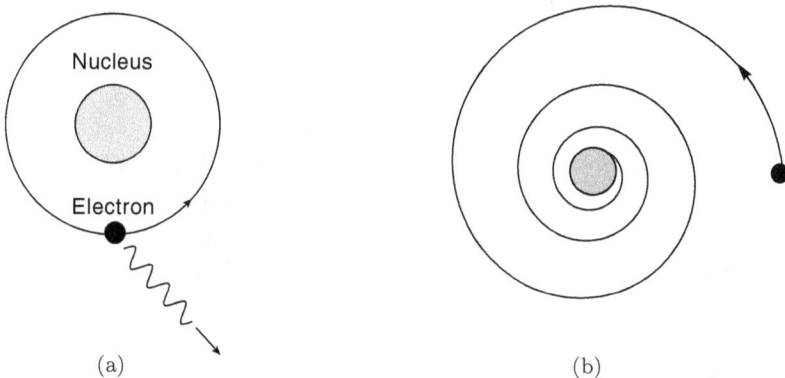

(a)   (b)

Fig. 5.1.   (a) An electron orbiting a nucleus radiates electromagnetic waves, losing energy in the process; (b) this must be compensated by a change in potential energy, which the electron can achieve by spiralling ever closer to the nucleus.

to be modified when objects move sufficiently fast. We might think that similar ideas will rescue us here: as the electrons spiral into the nucleus, they will travel faster and faster so that the effects of Special Relativity become more and more important. However, we saw that Special Relativity was needed to reconcile electromagnetism with the laws describing how objects move. Thus, the laws of motion had to change but not the laws of electromagnetism themselves. In order to explain how atoms can survive, we clearly need something else.

5.1.2 *The ultraviolet catastrophe*

Further evidence that something was deeply wrong with the laws of electromagnetism arrived in various forms in the late nineteenth century. One of these concerned what happens when you heat up objects. Generally this involves using electromagnetic waves to make the atoms inside the object jiggle around more, where more jiggling is what we mean by a higher temperature. In turn, hotter objects can themselves give off radiation, which you will know immediately if you accidentally touch a baking tray that you have just taken out of the oven. You will also know that hot pieces of metal can glow, which literally means that they radiate electromagnetic waves, at least some of which correspond to visible light.

In general, then, hot objects both *absorb* and *emit* radiation, and some combination of these effects is going on at any particular time. If an object's temperature is not changing, then it implies that there is just as much radiation being absorbed as emitted. An open problem at the end of the nineteenth century was to work out precisely what the *spectrum* of emitted radiation looked like. That is, given that electromagnetic waves can be classified according to their wavelength, we can ask how much radiation is emitted by a given hot object for each wavelength. The measured result is shown in Figure 5.2, as the solid line. We see that the amount of radiation becomes zero at either very low or very high wavelengths. In between, there is some peak value, which tells us that we "mostly" see radiation of a given wavelength. If this wavelength corresponds to visible light, it will constitute a particular colour. This then explains why hot objects can glow in different colours. It also tells us that, if the peak amount of radiation corresponds to a wavelength that is not in the visible part of the

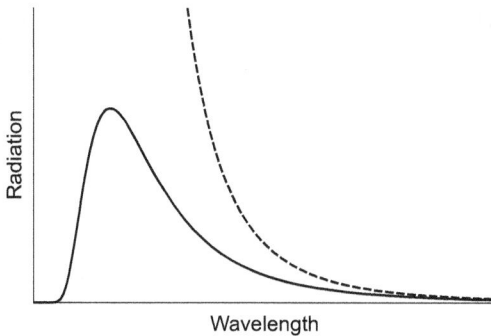

Fig. 5.2. Hot objects emit electromagnetic radiation of different wavelengths. The solid line shows the measured result for the amount of radiation emitted for each given wavelength. The dashed line shows the result one expects from the laws of electromagnetism.

electromagnetic spectrum, then we can still say the object is glowing. But this glow may only show up if we use a device that can detect non-visible light, e.g. an infrared camera. It is relatively easy to measure the spectrum shown in Figure 5.2. However, given that this corresponds to radiation of electromagnetic waves, it should be possible to calculate what the spectrum should be, as a direct consequence of the equations of electromagnetism. This calculation was carried out towards the turn of the twentieth century, and the result is shown as the dashed line in Figure 5.2. You do not have to be any kind of expert in electromagnetic theory to know that the dashed line looks nothing like the solid line! In fact, as a description of the latter, it is embarrassingly awful. It agrees with the measured result at large wavelengths but quickly starts to deviate it from it as we go to lower wavelengths. Even worse, the amount of radiation predicted from the equations goes up and up as we go to shorter wavelengths, in turn implying that any kind of hot object must emit *infinite* amounts of energy, of shorter and shorter wavelength. This is obviously false, and thus something very bad has gone wrong in the calculations. So bad was this problem for the physicists at the time that it was dubbed the *ultraviolet catastrophe*, where this name originated from the fact that ultraviolet radiation has a shorter wavelength than visible light, and it is precisely the short wavelength part of the spectrum that seems to be causing all the problems.

5.1.3   *The photoelectric effect*

As a final example of how electromagnetism must go wrong, we can consider an obscure property of metals known as the *photoelectric effect*. What distinguishes many metals from other types of solid materials is that some of the electrons in the metal are able to break free from their atoms and go wandering through the metal at large. Given that there is still some electric attraction between the electrons and the underlying network of atoms in the metal, however, the electrons are not usually allowed to leave the metal itself. A good analogy for this is to consider that you are walking as part of a large crowd in a busy shopping street. While you are free to walk among the various shops, you are not allowed to suddenly leave the crowd by floating up into the air!

We can change the situation in a metal by shining electromagnetic radiation on it. Radiation carries energy, and this energy can be absorbed by electrons, such that we can literally knock them out of the metal. To carry further our crowded street analogy from earlier, it is as if a strong gust of wind has suddenly hit the crowd, such that it lifts yourself – and possibly others – up into the air, leaving everyone else down below. Detailed experiments on metals have established properties such as how many electrons leave a given metal for a given amount of radiation and also how the energy of the emitted electrons depends on the wavelength of the radiation. But these experiments lead to some extremely puzzling features. To see why, let us remind ourselves that Maxwell's equations tell us that electromagnetic waves are smooth patterns of electric and magnetic fields that travel at a certain speed and can carry energy. The most plausible explanation for the photoelectric effect is that the energy in the waves is transferred to the electrons such that, once the latter have enough energy, they can leave the metal. If we take a very dim beam of radiation, with some correspondingly puny amount of energy, we would expect to have to wait a long time before the electrons could be emitted. Contrastingly, if we have a very bright beam of radiation carrying lots of energy, we would expect electrons to be emitted a lot more quickly. Furthermore, we would expect the emitted electrons to be travelling faster on average if we use a brighter beam: given that a brighter beam has more energy in it, and this energy is "given" to the emitted electrons, they should be more energetic and thus faster!

What actually happens is very different. First, while the number of emitted electrons indeed depends on how bright the beam of radiation is for some wavelengths, there is a maximum wavelength above which no electrons are ever emitted, *no matter how bright the beam is*. In our earlier analogy, it is as if there are certain types of winds that, no matter how strongly they blow, will always be incapable of lifting anything off the ground. Conversely, at other wavelengths, there are always electrons emitted, no matter how dim the beam. In the wind analogy, this is like having a very slight breeze, which nevertheless is capable of lifting people from the ground! Yet another puzzle is that, if we look at the speed of the fastest electrons that get knocked out of the metal, it does not depend on how bright the beam is, in direct contradiction with our above expectation. In the crowd analogy, this corresponds to saying that if a strong wind picks you off the ground with some speed, then a wind that is 100 times stronger makes you move no faster. Whatever is going on, it does not have anything to do with how waves are meant to work, electromagnetic or otherwise.

In this section, we have seen a number of different problems with electromagnetism, all of which involve particles of similar or smaller size than an atom. Any one of these problems would be difficult enough for the theory of electromagnetism to be able to cope with. That they all arrived at more or less the same time historically was a major source of confusion for the scientists of the time. We will see, however, that an amazingly straightforward idea comes to the rescue.

5.2   Photons

We have now seen three cases in which observed phenomena are apparently incompatible with the laws of electromagnetism. Furthermore, we can see that all of these phenomena involve electromagnetic waves. Atoms should collapse given that the electrons inside them radiate electromagnetic waves. The spectrum of radiation emitted by a hot object explicitly involves electromagnetic waves, as does irradiation of a metal so as to knock out its electrons. Thus, all clues are pointing towards the fact that we must modify the concept of a smooth electromagnetic wave that can carry any amount of energy it likes.

To see how this works, it is easiest to start with the photoelectric effect and to note that all of the puzzling features it implies can be gotten rid of if one merely assumes the following: *electromagnetic waves cannot carry any energy they like but must instead have some minimum unit, or "quantum" of energy, which can depend on the wavelength.* It will not be immediately obvious how this helps, so let's proceed carefully. First, the fact that radiation must come in little lumps tells us that the way the electrons leave the metal is that they must somehow absorb one of these lumps of energy. There will necessarily be some minimum amount of energy that each electron needs to be able to leave, and thus the amount of energy carried in a single lump must be above this minimum amount in order to knock out a given electron. For some wavelengths, the energy of each lump won't be high enough to allow the electrons to leave, and so they will remain stuck. Increasing the brightness of the radiation will not change matters: this simply amounts to throwing more lumps of radiation at the metal. If none of the lumps allows the electron to leave, throwing billions more at the electron will not change matters. The lump idea also explains the fact that the speed of the fastest electrons leaving the metal does not depend on the brightness of the beam: the energy of the emitted electrons depends on the energy of the lump of radiation they absorbed. If each electron absorbs a single lump before leaving the metal, throwing more lumps at the metal will not change the speed of the fastest electrons – only the number of electrons that get knocked out. Finally, we can also now explain the fact that for wavelengths at which electrons are emitted, this will happen even for a very dim beam of radiation. We are now assuming that our beam of radiation contains little lumps of radiation. A dim beam contains less of these lumps, but just one lump is enough to knock out an electron.

This simple idea is also enough to patch up the dreadful calculation of the radiation emitted by hot objects, that we saw in Figure 5.2. It turns out that the problem with the calculation using conventional electromagnetism is that it is assumed that electromagnetic waves can carry any energy. Now, however, we are saying that the wave is actually composed of little lumps of radiation. The fancy physics word for these lumps is *quanta* (the plural of *quantum*), and we say that the wave is *quantised*. This quantisation clearly implies that we must modify the calculation of the emitted radiation, where we must also include that the energy of each lump depends upon the wavelength in the way suggested by the photoelectric effect. It turns out that quanta of electromagnetic waves with *lower* wavelength have a *higher* energy. As the wavelength of emitted radiation decreases – corresponding to the left-hand side of the plot in Figure 5.2 – the energy of a hot object is simply not enough to be able to create the very high energy quanta corresponding to short wavelengths. Thus, as the wavelength of the emitted radiation decreases, it becomes impossible to emit such radiation, and this is clearly demonstrated by the solid line in Figure 5.2: the radiation indeed becomes zero as the wavelength goes to zero on the far left-hand side. A detailed calculation was done at the beginning of the twentieth century, and the result precisely matches the solid line in Figure 5.2! The fact that two entirely different issues were resolved by the same hypothesis – that electromagnetic waves are quantised – was very strong evidence for this hypothesis being correct.

How are we to interpret the fact that electromagnetic waves are quantised? In other words, what do the quanta correspond to, and can they be observed? The answer to this is very simple. To all intents and purposes, the quanta of the electromagnetic field behave like particles. If, for example, we shine a very dim beam of light onto some device that can detect it, we can measure the effect of each single quantum as it hits the detector. Similarly, the light particles can be thought of as travelling at a certain speed (the speed of light of course!) and also carrying energy by themselves. Whether they are "real" or not depends somewhat on your philosophical proclivities, but particles of light are certainly no less real than anything else we talk about in life, e.g. tables, clouds, cats, or ice creams. We then need a special name for particles of electromagnetic radiation that distinguishes them from the other types of fundamental particles that we have seen before. They are called *photons*, and the equations describing them make it clear that they originate from electromagnetic waves. Thus, there is a sense in which photons, while being particles, also somehow know that they are wave-like. Large numbers of photons, for example, will display wave-like properties when they interact, similar to how the peaks and troughs of water waves can cancel each other out. This causes a lot of confusion among beginning students in physics, who ask the very reasonable question: how do we know when to treat light as a wave and when to treat it as a particle? Much of this confusion stems from the fact that what we mean by a "particle" in this context is not the same as the particles that we first consider when learning physics. Our everyday intuition of how particles are meant to behave – which is itself rooted in observing large objects that obey Newton's laws – is simply not correct when thinking about photons.

At this stage, we have seen that the electromagnetic field is not completely smooth but instead has to be quantised. In Chapter 2, we saw that there is a basic distinction in the world, between matter on the one hand (the "stuff" that things are made of) and forces (that tell the "stuff" how to move about). In Chapter 3, we saw that the electromagnetic field is where electric and magnetic forces come from. Thus, quantisation of electromagnetic waves tells us that forces are somehow quantised, and we formalise this idea in the following chapter. Before doing so, it is natural to ask whether the quantum idea extends to matter as well. Indeed it does, as we now explore.

5.3   Matter Is Quantum Too

The fact that light is composed of individual photons explained two out of the three problems with electromagnetism that we saw in Section 5.1: how light knocks electrons out of metals (the photoelectric effect), and how radiation is emitted by hot objects. Arguably the most serious and dramatic problem in Section 5.1, however, was that atoms cannot apparently exist but instead must decay almost instantly due to the fact that the orbiting electrons emit large amounts of radiation before spiralling into the nucleus. Given that the stuff around us is indeed made of apparently stable atoms, how can this be explained?

The answer is to use a similar concept to the photon idea. We saw that individual photons (of a certain wavelength) carry a set amount of energy. Another way of saying this is that the energy of an electromagnetic wave cannot be anything it likes but must instead come in certain fixed amounts, corresponding to different numbers of photons. Let us now apply something like this to the electrons inside an atom where, for simplicity, we can take the simplest atom there is. This will have a single proton in its central nucleus and a single electron orbiting around the outside. We need a charged particle at the centre to attract the electron, hence why we cannot use a neutron for the nucleus. Furthermore, the fact that there is an equal number of protons and electrons (only one of each) implies that the atom is electrically neutral overall. The name for this atom – as it occurs in the famous periodic table of the elements – is *hydrogen*, and merely mentioning its name will strike pangs of fear into any third-year university physics student, who typically has to examine the mathematics of how it works in full gory detail. Here we can avoid this complication and simply look at what was ultimately going wrong in Figure 5.1: given that the electron could have any amount of energy, it could continuously lose this energy by radiating it away in the form of electromagnetic waves. This is what in principle causes the electron to spiral into the nucleus, and it won't be able to happen if there is instead a fixed set of distinct energies that the electron can have. In particular, we will assume that there is some minimum amount of energy the electron is allowed to have and such that it is, on average, an appreciable distance away from the nucleus. As soon as we stipulate this, collapse of the atom becomes impossible: the closest the electron can be to the nucleus is to occupy the state of the atom that has the lowest possible energy, by definition. Then there may be higher allowed energies, such that the electron may be further away. The set of allowed energy levels (in some suitable units of measurement) are shown in Figure 5.3. Increasing energy is shown on the vertical axis, and I have calculated and plotted what our current theories of the hydrogen atom tell us the allowed values of the electron energy are. We see that there is indeed a fixed set of values and that there is a minimum energy. Further higher energy levels will exist, which are not shown.

So far we have invoked the idea of quantisation of the electron energy in order to explain why atoms can exist. But there is a very nice experiment that we can do using real atoms that immediately tells us that this is in fact true. Imagine that we start with a hydrogen atom in which the electron has the lowest possible energy. Now consider shining a beam of light on the hydrogen atom, where we can use white light consisting of all possible wavelengths. The different wavelengths correspond to different colours of visible light, and the usual way to talk about wavelengths is to use nanometres,

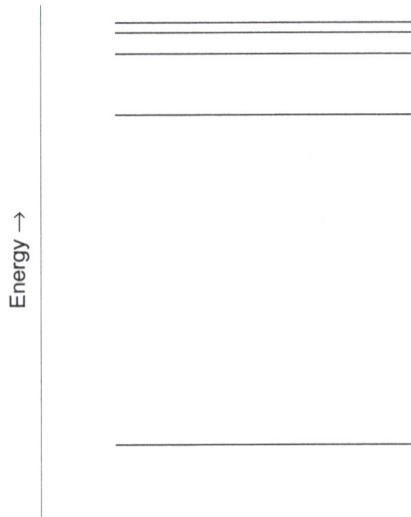

Fig. 5.3. The electron in a hydrogen atom cannot have any energy but instead only energies from a certain fixed set, which contains a minimum allowed value.

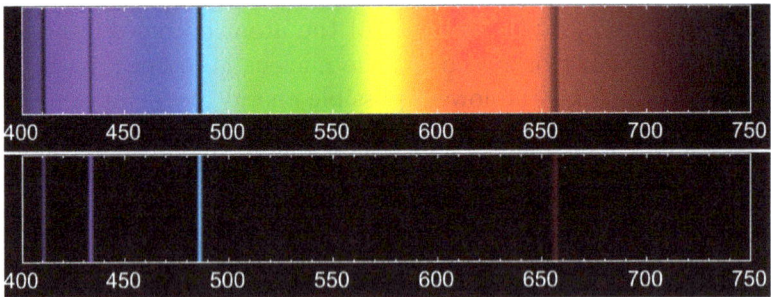

Fig. 5.4. Upper panel: Radiation absorbed by a stationary hydrogen atom cannot have any wavelength but instead a choice from a given fixed set. This is directly relatable to the fixed energy levels in Figure 5.3. Lower panel: Likewise for the radiation *emitted* by a hydrogen atom.

where 1 nm is 0.000000001 m. Then we can associate a number with each possible wavelength, which will in turn correspond to a colour. Something like this is shown in the upper panel of Figure 5.4, which shows the possible colours of visible light and an accompanying scale of different wavelengths. If you don't understand what the numbers mean, there is no need to worry, as they will not be needed for the following discussion.

If we shine our beam of white light on the hydrogen atom, the electron may absorb some energy from the light. However, Figure 5.3 implies that it cannot absorb any amount of energy we choose. There are only certain allowed energies that the electron can have, and it must therefore absorb precisely the right amount of energy so that it moves from the lowest energy value in Figure 5.3 to one of the others. How this happens may or may not already be clear, but let me spell it out. What we know from the previous section is that the light beam itself consists of lots of photons, each of which has a particular energy that depends upon its wavelength. Thus, there is a straightforward way in which the electron can absorb a given amount of energy: it can somehow absorb a photon, whose energy is precisely such as to correspond to the difference between two of the energy levels in Figure 5.3. A given spacing between two energy levels fixes the energy of the photon that must be absorbed, and this in turn fixes the wavelength. It follows that, if we shine a beam of white light on a hydrogen atom, there are only certain wavelengths of radiation that the atom can absorb. Namely, those wavelengths whose corresponding photons have energies that allow the electron to jump into an allowed energy level.

The earlier discussion is rather convoluted, but the upshot is straightforward: if we shine white light on some hydrogen atoms and then measure the light coming out the other side, we should see that certain wavelengths are missing. These correspond to the allowed wavelengths of light that can be – and have been – absorbed by the atom. We can literally see this in the upper panel of Figure 5.4, which shows the result of such an experiment. We see almost all of the colours of visible light, but there are four dark lines, constituting the "missing" wavelengths, whose photons have been absorbed by the hydrogen atom.

In our earlier discussion, we started with a hydrogen atom whose electron had its lowest possible value of the energy. However, we could instead have started with a hydrogen atom whose electron already had a higher value of the energy, provided this is one of the allowed values shown in Figure 5.3 (or one of the others that exists but not shown explicitly). This has the wonderful name of an *excited hydrogen atom*, and it will become less excited – or not excited at all – if the electron drops down to a lower energy level, thereby emitting a photon of light in the process. Similar to the discussion of absorption of a photon, the photons emitted by the atom must have very particular wavelengths, again corresponding to the allowed spacings between the energy levels in Figure 5.3. The various emitted colours of visible light are shown in the lower panel of Figure 5.4, and we see that these precisely complement the missing lines in the upper panel, as they should: the possible wavelengths of light that can be absorbed by the atom must precisely match the wavelengths that can be emitted, as both are consequences of the fixed energy levels in Figure 5.3.

Although we have talked about the hydrogen atom here, similar ideas apply to other types of atoms. That is, there is always a set of allowed energy levels that each electron can have, and this will in turn imply that only certain wavelengths of light can be emitted, which may include electromagnetic radiation that is outside the visible part of the electromagnetic spectrum (as indeed happens for hydrogen, corresponding to lines not shown in Figure 5.4). This is itself highly useful: the set of measured colours constitutes a sort of fingerprint that uniquely distinguishes one type of atom from another. It is this fact, for example, that allows us to know what kinds of atoms are present in distant stars and galaxies: if we can measure the light that reaches us from those galaxies, then we can see which wavelengths are missing and thus work out which atoms are present. It is the quantisation of electron energy that makes this possible.

We have by now made clear that quantising the energy of electrons in atoms resolves the puzzle of why atoms can exist in the first place and also that experiments tell us that this is true. However, this is very different to having a detailed theory that correctly includes the fact that electrons can only have fixed energy values. In principle, this theory should allow us to calculate the different energy values, for different types of atoms. Going further than this, the theory should correctly explain the properties of matter particles in all situations, not just those in which electrons are confined in atoms. Given that this theory must build in the effect of quantised energy, it is known as *quantum mechanics* and took literally decades to understand in the first half of the twentieth century. The consequences of the theory are so far-reaching and bizarre that they are still being discussed by philosophers today. But it is indeed true that the laws of quantum mechanics must replace those of Newtonian mechanics, at very small distance scales.

5.4 Quantum Mechanics

To understand why a theory of quantum mechanics might be difficult, we can start by thinking about how we describe things that we
measure in everyday life. If you are buying a bookcase, and also being
sensible about it, you will probably do some measurements first to
make sure that the bookcase will fit in your intended location. This
involves getting a measuring tape or ruler and associating a number with the length you are measuring, which may be in centime-
tres, inches, or some other units. You will also be used to measuring
times in seconds, minutes, hours, months, years, or decades. As well
as lengths and times, there are many other things in life that we
can put numbers on, such as weights, angles, speeds, temperatures,
areas, and volumes. In all of these cases, there is a single number that
describes the quantity involved and where this number may take any
value within reason (e.g. some numbers can be negative or positive,
whereas others must be positive). We have also seen quantities – such
as the velocities discussed in Chapter 2 – where there is a direction
associated with the quantity, as well as a size. However, also in these
cases the size of the quantity is a simple number that can again take
any reasonable value. So widespread is our use of numbers in everyday
life that we may have barely stopped to notice that they are actually quite an abstract mathematical object by themselves. In fact,
there is a whole branch of mathematics – known as *number theory* –
that studies the various properties that numbers can have and whose
non-obvious applications include the highly important goal of making
sure our data are secure when transported over the internet.

As soon as we say that numbers are a certain type of mathematical
object, it implies that there are other more complicated mathematical objects, whose properties will be different. Indeed there are such
objects, and we can then ask the following question: if we want to
write a theory in which energy and other quantities can only take
certain fixed values, what types of mathematical objects should we
use as the basis of our theory? The answer to this involves a level of
mathematical abstraction that takes several years of university education to appreciate, so I will not answer it here. However, a point
that I think is easier to realise is that in our quantum theory, we cannot simply represent quantities such as energy and velocity by simple
numbers. These must necessarily take a smooth range of values, and

we would need some weird mechanism for saying why almost all values except a special restricted set are somehow not allowed. Instead, we must use mathematical objects which have a fixed set or "spectrum" of allowed values built in. How to do this was a major puzzle historically and explains why it took decades for a full theory of quantum mechanics to be established. Part of this involved the fact that there were different rival versions of the theory that were ultimately realised to be equivalent to each other once the mathematics was fully understood.

We have seen something like this situation before. When the need for Special Relativity arose, Einstein and others started with the requirements that the theory had to have (i.e. constancy of the speed of light and the principle of relativity) and used these underlying principles to dictate what the final theory should look like. This theory had very strange consequences, such as time dilation and length contraction, that did not make any sense at first. However, the correct thing to do was to trust the mathematics guiding the theory and to then check that the final theory made sense using detailed experiments, rather than our (imperfect) intuition. Quantum mechanics is also a theory that is "designed" to incorporate well-observed features, namely that energy and other quantities should be quantised. In doing so, however, it forces us to use more complicated and abstract mathematics than we are used to. Furthermore, when the final theory takes shape, its consequences are even more shattering to our everyday notions of how the world behaves than Special Relativity ever was. As we will see, quantum mechanics even compels us to redefine what we mean by science itself!

5.5   Matter Particles Are Waves

In order to understand the implications of quantum mechanics in more detail, it helps focus on a particular formulation of the theory that was first developed by Schrödinger and others in the early twentieth century. The key idea was to change our notion of what an electron is. Before the birth of quantum mechanics, it was assumed that electrons were well-defined particles, with a particular location. Schrödinger's equations of quantum mechanics instead posit that the electron can behave like a wave, in a similar way to other types of waves in physics. We interpret these waves more fully later on, but let us first see why describing electrons as waves indeed allows us to understand how the energy levels of electrons in atoms can only belong to a restricted set of values. So far, we have seen examples of waves – such as electromagnetic waves – that are travelling along without stopping. A similar situation occurs for the water waves that you see in the ocean. If you are far away from any land, the water waves will travel ever onwards, eventually stopping when they see a beach. This is very different, however, to water waves in a bowl or sink. If you drop your hand in a bowl of water, you will create waves that make the surface of the water bob up and down, but which cannot travel out of the bowl, and thus do not move horizontally in any particular direction. They are called *standing waves* and are essentially what happens if waves get confined to a particular region of space (in this case the bowl).

Another familiar example of standing waves in everyday life is the vibrations that you can see on a guitar string, as shown in Figure 5.5. These vibrations cause the air to vibrate inside and around the guitar, causing sound waves to propagate to your ears. But if we look at the string itself, we notice something interesting. There is the fact, for example, that a given string is clamped at either end so that it cannot move there. Then, were we to zoom in and pluck the

Fig. 5.5. A set of vibrating guitar strings.
Source: Photo by Daniel G on Unsplash.

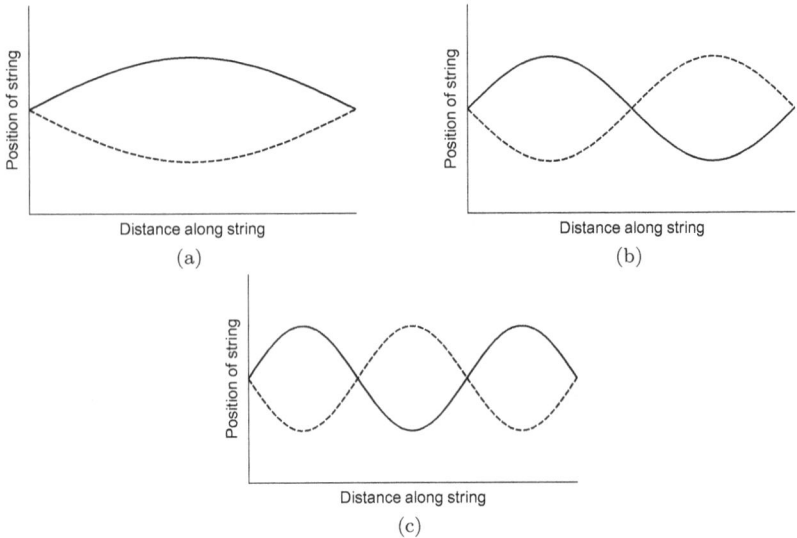

Fig. 5.6. Different patterns of vibration of a guitar string, where the solid line shows the profile of the guitar string at a given time, and the dashed line the profile at a later time. Pattern (a) requires least energy, and patterns (b) and (c) require more energy to happen.

string in different ways, we would see that there are *different* ways in which the string can vibrate and that in general some combination of these is going on at any one time. Examples of the different patterns of vibration are shown in Figure 5.6, and we can interpret these pictures as follows. Looking at Figure 5.6(a), what I have shown by the solid line is what the string is doing at some given time, where the horizontal axis represents the string if it is not vibrating, and the vertical direction in the figure represents how the string has been displaced by its vibration. We see in particular that the string has no displacement at the beginning and ends of the string, corresponding to the fact that the guitar string is clamped there and so cannot move. We have also shown a dashed line representing what the string is doing at some later time. At in-between times, it is oscillating between these two positions, although this usually happens so fast (hundreds of times per second) that it is impossible to see with the naked eye.

Figures 5.6(b) and 5.6(c) show alternative patterns of vibration, in which more is happening between the end points. We see that there are now positions on the string where there is no motion at all, at any time. These are called *nodes* and are a common feature of standing waves. Between two neighbouring nodes (including the endpoints), we see a similar wave pattern to that of Figure 5.6(a) but compressed so that it fits in the right gap. Looking at the three different pictures in Figure 5.6, you may be able to guess what the next few patterns in the sequence look like. They will have more nodes and thus more "complete" wave patterns along the string. Indeed, there is a well-known equation in physics – typically covered in the first year of a university degree – whose solutions tell us what these distinct wave patterns should be and also proves for us that any general vibration of the string will consist of some combination of all of these distinct patterns. Furthermore, the equation tells us that vibration patterns that have more nodes require more energy to occur. That is, we must pluck the string harder in order to create the more complicated patterns, as perhaps makes sense given that more is going on.

What does any of this have to do with electrons in atoms? The point that the guitar string illustrates is that if we trap or confine waves, they cannot have any energy they like but must have a fixed set of specific energies. In the guitar example, these are the energies corresponding to each of the distinct patterns of vibration. As soon as we say that, (i) electrons are waves and (ii) they are confined within atoms (which they are by the electromagnetic force between the electrons and protons), then it follows *automatically* that their energy must be quantised! Of course, the patterns of vibration that electrons can have in atoms are more complicated than the vibrations of a guitar string. They are described by a different equation for a start, and we must also worry about all three dimensions of space. Nevertheless, we can still visualise the different vibrations of our electron waves. I have plotted some of them in Figure 5.7, where the darker colours represent a greater "size" of the electron wave at that point. These shapes are known to chemists as *atomic orbitals*, given that they represent – in some very abstract and fuzzy sense – the orbits that electrons can have in atoms. As in the case of the guitar string, the more complicated orbitals correspond to higher energy vibrations of the electron so that the set of fixed electron energy levels arises from the fact that the equations predict certain orbitals and not others.

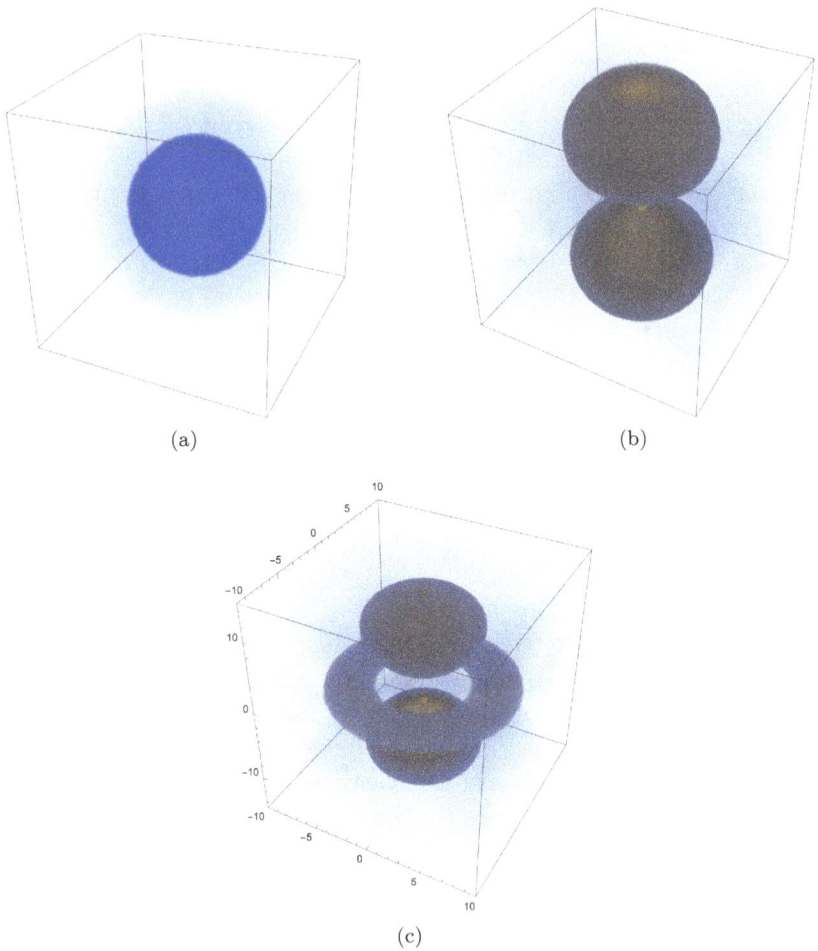

Fig. 5.7. Visualisation of the different ways in which an electron wave can vibrate inside an atom. The darker the colour, the larger the size of the wave at each point.

So far so good. But how are we meant to interpret what is meant by an "electron wave", when this was meant to be a particle? In the case of the guitar string, we knew what it was that was vibrating in a wave-like way, but it is not at all clear what this means for an electron. Indeed, this question caused a great deal of bother when quantum mechanics was first being understood, and one of the most common interpretations of the theory – backed up by detailed experiments over many years – starts by making a striking claim about our relationship with the world around us. In conventional physics up to this point, which includes the theory of Special Relativity in the previous chapter, we could always talk about objects that had a precise location. It might be that our measuring devices are imprecise or that we only need to know the rough location of something for any given purpose. Also, an object may be moving, in which case its position will change over time. But in principle, we could consider a hypothetical perfect measuring device that would give the *exact* location of an object at any given moment.

In quantum mechanics, this is not true. The way to interpret what our electron waves mean is that they tell us, at any given point in space, *how likely it is* that we will find our electron there. That is, the darker colours in Figure 5.7 represent places where we are more likely to see the electron. This may not sound very serious, but it is a profound difference to the science that came before that should shock us all. As soon as we say that the theory can only ever predict the chance, or probability, that the electron should be in a given place, we have fundamentally changed what we mean by science. Before quantum mechanics, people thought that if we knew the precise state of the universe at any given time, we could uniquely predict what happens in the future. Quantum mechanics destroys this hope and, in doing so, changes our relationship with the reality around us. In other words, science allows the possibility that different outcomes are inevitably possible and that something (or someone) has to choose between them. Philosophers are still arguing about what this might mean. Perhaps surprisingly, though, it has never prevented us from actually calculating things in physics and making new technologies.

If the electron could potentially be in many different places, what happens when we actually measure an electron? We can certainly place detectors in specific locations that go beep when an electron hits them. What has happened in this case is that our predicted guess of where the electron might be suddenly collapses, and we find the electron in our detector. We have here talked about the position of the electron, but similar considerations apply to other aspects of electron behaviour, e.g. energy, velocity, and so on. If we pick a specific thing we could measure – e.g. the energy of an electron in an atom – then at any given time there is no well-defined value of this quantity. Rather, we should think of the atom we are watching as in some sort of weird superposition or combination of states of different energy. As soon as we decide to measure the energy, however, the system collapses into a single state of well-defined energy. By merely observing the world, we irrevocably change it, and this extremely strange idea also runs counter to our traditional notion of science, whereby the world proceeds merrily by itself, whether or not we choose to watch it. It also raises the thorny question of what constitutes an observer. If we place a device for measuring electron energies in a locked room and leave it, does the state of the atom collapse? Or does this happen only if we are explicitly watching the detector? Some philosophers and scientists have even speculated that quantum mechanics may be tied up with our own consciousness, in that the latter would provide a mechanism for insisting that the state of the world suddenly changes. But how can it do this without our knowing what is going on? Another interpretation of quantum mechanics – which mathematically can be taken just as seriously as the traditional interpretation – is that every time a measurement or observation is made, the universe splits into multiple different universes so that all of the myriad possibilities are happening in some universe somewhere. This may sound like the stuff of science fiction, but the equations demand to be taken very seriously.

5.6 Is Quantum Mechanics Enough?

In this and the previous chapter, we have seen that new theories are needed to replace the traditional Newtonian view of the world. One of these theories – Special Relativity – arose when objects move very fast so that their speeds approach that of light. The other – Quantum Mechanics – is needed when objects get very small. We can think of collecting all these theories together, and I show such a scheme in Table 5.1. It is hopefully immediately obvious that this is a very useful exercise to do. If we look at the table, we see that there is a missing section in the lower-right corner, corresponding to a theory that should describe very small things that are moving very fast. Such objects certainly occur in the universe: high energy particles are hitting the Earth from space every second. On Earth itself, subatomic particles are routinely accelerated to very high speeds at particle accelerator experiments, such as the Large Hadron Collider at CERN. Thus, we clearly need to be able to describe such situations.

Table 5.1. The range of physical theories
applying to different physical situations.

		Speed of Object	
		Slow	Fast
Size of Object	Big	Newtonian Mechanics	Special Relativity
	Small	Quantum Mechanics	?

On a more fundamental level, we have seen in the cases of both Special Relativity and Quantum Mechanics that these theories effectively replace Newtonian mechanics, providing a more underlying and correct description, that must somehow reduce to Newtonian mechanics in an appropriate limit. The missing theory in Table 5.1 occupies the unique position of being the most fundamental theory that in turn must reduce to any of the other theories as objects become larger or slower. It is thus the theory that gets us closer than any other in describing the basic building blocks of nature, and we examine its intricacies in the following chapter.

Summary

In this chapter, we have seen the following ideas:

- The standard theory of electromagnetism leads to inconsistencies, such as the fact that atoms cannot exist.
- This is cured by a theory in which electromagnetic waves are composed of lumps, or quanta, called *photons*.
- Electrons behave like waves. When these get trapped in atoms, the energy becomes quantised due to only certain patterns of vibration being possible.
- The corresponding theory is called *Quantum Mechanics* and tells us that there are no certainties in nature. Instead, we can only predict the likelihood of various outcomes.
- When we perform measurements, one of the possible values of what we are measuring gets somehow picked out.
- If things are small but also move very fast, a new theory is needed, replacing both Special Relativity and Quantum Mechanics.

Chapter 6

Everything Is Made of Fields

In the last chapter, we saw how quantum mechanics modifies the behaviour of light waves, which become particle-like in the form of photons. We also saw that quantum behaviour can make particles like the electron behave like waves. While these ideas sound similar, it is perhaps unsatisfactory that they are not quite the same. Early on in this book, we saw that there appears to be a basic distinction in nature between matter and forces. Particles like the electron are examples of what we mean by matter, and electromagnetic waves are examples of the fields that give rise to forces. Thus, the two types of quantum behaviours that we saw in the previous chapter seem to imply that forces and matter are to be somehow treated differently. There is nothing especially wrong with this, but wouldn't it be nice if both matter and forces arose from the *same* type of theory?

We also saw, in Table 5.1, that there is apparently at least one type of theory that is missing in everything we have seen so far, namely one that successfully combines the effects of Quantum Mechanics and Special Relativity at the same time. Such a theory is needed to describe small objects (e.g. subatomic particles) that are moving at speeds close to that of light, and indeed such a theory exists. It is called *Quantum Field Theory*, and we have in fact already seen its main ideas when we talked about photons. We will shortly see how this theory applies to both matter and forces, but let us first note that Table 5.1 can now be completed, as shown in Table 6.1.

Table 6.1.   The range of physical theories applying to different physical situations.

		Speed of Object	
		Slow	Fast
Size of Object	Big	Newtonian Mechanics	Special Relativity
	Small	Quantum Mechanics	**Quantum Field Theory**

Quantum Field Theory includes the effects of both
Quantum Mechanics and Special Relativity.

6.1   Quantum Field Theory

In Chapter 3, we saw that electric and magnetic forces are associated with electric and magnetic fields, where the definition of a field is some quantity that is defined at all points in space and at all times. The electric and magnetic fields certainly fulfil this criterion: we saw in Figures 3.3 and 3.5 that we can represent these fields as arrows at each point in space. As time evolves, these arrows might also change in general. Whenever we have a field in nature, we must have some equations that describe this field. In the case of electromagnetism, these are the Maxwell equations discussed in Chapter 3. They describe, in full generality, precisely how electric and magnetic fields are generated by charged particles, be they moving or stationary. We also saw in Figure 3.6 that particular solutions of Maxwell's equations are such that they describe waves of electric and magnetic fields that travel through space at the speed of light. Indeed, these waves correspond to light itself, plus other kinds of electromagnetic waves that go by different names (e.g. X-rays and microwaves).

In Chapter 5, we saw that quantum mechanics modifies the wave-like nature of electromagnetic waves. Rather than having any old energy, waves of a given wavelength consist of individual lumps or *quanta*, where these can be thought of as corresponding to particles called *photons*. It is common to group together the electric and magnetic fields into a single field, which contains both effects, and the physicists usually call this the *photon field*, in anticipation of the fact that it gives rise to photons upon "quantising" the theory of electromagnetism. Interestingly, this single field can give rise to any number of individual photons. Thus, our quantum theory of electromagnetism is a theory of many particles, not just one. However, it has only a single *type* of particle – the photon.

Given that the quantum theory of electromagnetism involves a field, and is also quantum, it makes perfect sense to refer to it as a *Quantum Field Theory*, or a QFT for short. Indeed, one way of seeing where QFT comes from is that it arises as the answer to the following question: what happens if I take a theory containing a field and make it quantum? That this includes both the effects of special relativity and quantum mechanics in this case is because photons travel at the speed of light. Thus, the non-quantum version of electromagnetism already has Special Relativity built in. Indeed, a property of electromagnetic waves (their constant speed) entered the very foundations of Special Relativity, as we saw in Section 4.2.

It may not surprise you to know that QFT has a formidable reputation as a highly difficult subject. It is typically first encountered in the fourth year of a university degree and even then is taken by only a small subset of students, many of whom are put off by the pages of algebra that inevitably occur. However, it is easy to see why the subject occurs only later on in a physicist's training: if QFT combines Quantum Mechanics and Special Relativity, it can only be learned after both of those subjects have been mastered. It is also important to realise – and this is a point that is seldom stressed in even the best courses on the subject – that the vast amounts of algebra needed to formulate QFT are in fact hiding relatively simple ideas that are learned right at the beginning of a physics education. At the heart of QFT are the following simple statements: (i) fields are described by equations, (ii) these equations have wave-like solutions, and (iii) quanta of these waves are particles. A summary of this for the electromagnetic case is shown in Figure 6.1, and it turns out that these basic ideas are far more general than they might first appear.

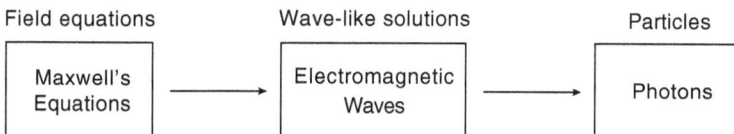

Fig. 6.1.   Any given field in nature is described by certain equations, which typically have wave-like solutions. In the quantum theory, these waves come in well-defined lumps or quanta. This is the central idea of QFT and is summarised here for the case of electromagnetism.

6.2   Matter Fields

When we looked at quantum electrons in the previous chapter, we saw that quantum mechanics describes them as waves, where the "size" of the wave represents how likely it is for the electron to be in a particular place. If we have many electrons, we can think of combining them all into a single wave-like thing that describes where all of the electrons are all at once. Indeed, this is what physicists do when they describe what happens in solid materials like metals, each of which contains a (very!) large number of electrons. However, for such a theory to make sense, it must be the case that the number of electrons in our system – or the universe as a whole – is not changing. After all, our waves are meant to represent where the electrons are. The size of the wave may be zero somewhere – telling us that an electron definitely won't be found there – but the whole wave theory has built into it that each electron must be *somewhere* and cannot suddenly decide to disappear.

Of course, there are very sound reasons to expect that electrons shouldn't be allowed to vanish. We have seen that they carry electric charge, and we also made reference earlier to the fact that electric charge in our universe is *conserved*, meaning that in any process in which particles interact and/or break up, the total amount of charge beforehand must be the same as the total amount of charge afterwards. If an electron were to suddenly disappear, it would take its charge with it, which would be a clear violation of the conservation of charge rule. Before we conclude that this is impossible, however, we should bear in mind the moral of this book so far: do not trust anything that appears to make sense! There is in fact a subtle loophole in the above argument, involving a particularly exotic feature of QFT.

6.2.1   *Antimatter*

It is possible to write an equation describing the electron, which includes the effects of Special Relativity. This was first done in the 1930s by Dirac, although his original understanding of what the equation meant has been modified somewhat since then. Nevertheless, we still refer to the equation describing the electron as the *Dirac equation*, and it has the puzzling feature that it seems to contain too many solutions. There are solutions that indeed represent the electron, corresponding to a particle with a certain mass and negative electric charge. As we saw in Chapter 4, particles can also have kinetic energy, associated with how fast they are moving. It turns out that this kinetic energy must always be positive, and the electron solutions of the Dirac equation indeed have this built in.

Alongside the electron solutions, there are other solutions of the equation that, puzzlingly, have a *negative* energy but otherwise look like perfectly normal electrons. The obvious thing is simply to say that such solutions must somehow be thrown away, as being not acceptable or valid in nature. However, this turns out to be inconsistent in the quantum theory. We saw in the previous chapter that electrons can fall from higher energy levels to lower energy levels, emitting a photon in the process. If we interpret the negative energy states as belonging to electrons – however hypothetically – then any given electron in the universe could continually lower its energy by dropping down to progressively lower energies, emitting infinite amounts of electromagnetic radiation as it does so. This is clearly nonsense and held back interpretation of the Dirac equation for some time. The right way of thinking is simple but very interesting: it is possible to reinterpret the *negative* energy electron states as belonging to *positive* energy particles, provided the latter are not electrons. Instead, they correspond to particles with the same mass as the electron but with an opposite electric charge. That is, whereas electrons are negatively charged, the new particles are positively charged, but where the amount of electric charge is the same in each case. These particles are called *positrons* or *antielectrons*. When Dirac first wrote his equation, it would have been a very bold claim to say that a new type of particle has to exist, merely to patch up how we interpret an equation. Remarkably, however, positrons were discovered a short time afterwards so that they are physical particles that actually exist.

The Dirac equation is also used to describe other fundamental particles in nature, such as the muon and tauon that are closely related to the electron, the quarks that live inside the protons and neutrons found in atomic nuclei, and the neutrinos that we saw in Chapter 3. It immediately follows that all of these fundamental particles have antiparticle equivalents, which are almost identical but have opposite electric charge where appropriate. The general name for such particles – to distinguish them from the matter particles we started with – is *antimatter*. These names, however, are a bit misleading. They suggest, for example, that matter is somehow more "fundamental" than antimatter, when in fact both of them are on a perfectly equal footing. There is nothing in Dirac's equation that tells us that either matter or antimatter is in any way special, and thus the names could just as easily have been the other way around.

The equations also give us a rather striking consequence: when matter and antimatter collide, they can annihilate each other in a burst of energy, leaving only electromagnetic (or other) radiation. The reverse process can also happen. That is, electromagnetic radiation can suddenly split into equal numbers of matter and antimatter particles. This situation crucially relies on a special feature of Special Relativity that we saw in Chapter 4, namely that mass and energy are ultimately equivalent. If energy becomes (anti)matter, the energy of the radiation is being somehow converted into the mass of the matter and antimatter particles, and vice versa. This may sound like the stuff of science fiction, but it has very real and life-saving applications. Modern hospitals routinely use a type of scan known as a PET scan, where the acronym stands for Positron Emission Tomography. In such a scan, the patient is injected with a small amount of radioactive substance that prefers to sit in certain locations in the human body (e.g. cancer cells, bones, or the brain). Atoms in this substance can decay into other atoms, emitting a positron in the process. Each such positron annihilates an electron in your body, and the resulting burst of energy can be detected to give a high-precision image of what is happening inside yourself. Although this sounds dramatic, the quantities of radiation are sufficiently low so as to be of minimal risk to the patient.

Note that when an electron and a positron (antielectron) annihilate each other, the number of electrons in the universe goes down by one, and an electron has genuinely disappeared. It does so, however, by eloping with a positron, such that the total electric charge in the universe – positive and negative for the positron and electron respectively – remains the same. Thus, it is perfectly possible for electrons to be able to disappear, and yet it still be true that charge is conserved. More importantly, the fact that electrons can indeed disappear tells us that the original quantum theory of the electron – in which electrons were not allowed to disappear – cannot be all there is. We must somehow be able to interpret Dirac's equation for the electron in a way that accurately represents what electrons and positrons get up to.

6.2.2 Fields for everything

Now that we know about antimatter, I can tell you the correct interpretation of Dirac's equation for the electron and other matter particles. It describes a field filling all space, just as Maxwell's equations describe electric and magnetic fields. The nature of the field (e.g. what kind of mathematical quantity it is) will not bother us here. Merely knowing that the electron has a field associated with it allows us to apply the scheme of Figure 6.1. Given the electron field, we can ask what the corresponding equation is. As stated earlier, this is the Dirac equation, which describes both electrons and positrons. We can then ask if this equation has wave-like solutions, and indeed it does. Finally, then, we can assume that these wave-like solutions have quanta associated with them, which behave like particles. Once again, this is indeed what the formalism of QFT, as applied to the electron, does for us. The interpretation of these particles are that they correspond to the electron and positron. What's more, just as the electromagnetic field could describe any number of photons, the electron field can describe any number of electrons and positrons. This scheme is summarised in Figure 6.2, which can be compared directly with Figure 6.1.

Applying this same idea to each of the fundamental particles in turn, we find that all matter in the universe is made of fields, where there is one field for every type of matter particle (plus its antiparticle). There are quark fields, for example, which describe both quarks and antiquarks. There is a muon field, for the muon and antimuon, and so on. Every time we have such a field, we can write down an equation describing it, solve it to get waves, and then say that the particles we know and love arise as quanta of these waves. It is worth

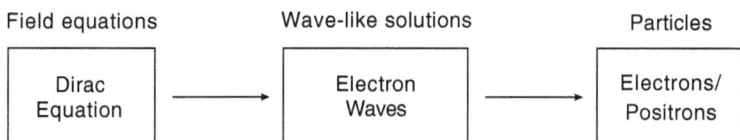

Field equations	Wave-like solutions	Particles
Dirac Equation	Electron Waves	Electrons/ Positrons

Fig. 6.2.   Similar scheme to Figure 6.1 in the case of the electron.

taking a few moments to properly digest this. What I am saying is that everything we see around us is made of particles, *all of which* arise from abstract fields filling all space. If we accept this, then there is not really any such thing as separate particles at all. All electrons in the universe come from a single electron field, as do all positrons. All of the quarks that make up atomic nuclei come from quark fields, and so on. This dramatically simplifies the nature of the universe. Before we knew about QFT, we would have wondered about where all the individual electrons came from, and all the separate quarks, etc. Now we realise that simply knowing about the different types of particle is enough: once we have a field equation for each of them, it will predict the existence of as many (anti)matter particles as we like. QFT also resolves a slight problem with the ideas of the previous chapter. There, we saw that electromagnetic fields seemed to behave differently to matter particles. Now, however, we see that *all particles*, be they associated with matter or forces, arise from the same type of physical object, namely a field. There remains a certain distinction between matter and forces in today's quantum field theories, but only because the mathematical properties of each type of field can be slightly different. The concept of the field itself is universal and means that quantum field theory is conceptually the most natural and elegant of the theories in Table 6.1, from which all other theories must ultimately follow.

As may or may not be clear, a QFT is a *type* of theory, rather than being a specific theory. We can construct different quantum field theories by putting in different types of matter and force particles. However, our own universe will have a given set of such things, and it is a natural question to ask: of all the possible QFTs, which is the one that we actually see in our own world? We have already seen one of the forces (electromagnetism), as well as the various matter particles that we must include (electrons, muons, tauons, neutrinos, and quarks). To complete the description, we need to talk about the other forces that exist.

6.3   Nuclear Forces

In Chapter 3, we focused on a particular force that can act on particles, namely that of electromagnetism. This is not the only force in nature. Another one that is very familiar from everyday life is gravity, which acts between massive objects to try to pull them together. We examine gravity in much more detail in the following chapter, but we can also ask what other forces there are in nature, in a similar way to how we can characterise all the different types of matter particles. It turns out that, although there are many different types of matter particles, there are only four distinct forces. We refer to these as *fundamental forces*, given that they underly all observed forces in our universe. The two fundamental forces mentioned earlier (electromagnetism and gravity) are familiar mainly because they are obviously visible in our everyday lives. We are all conscious of the fact that we are pulled towards the ground (gravity). Likewise, when we touch, pull, or push things, we become aware of electromagnetism, even if the precise details of what it is called or how it works might remain mysterious.

This leaves two fundamental forces unaccounted for, and they show up when we look at how atomic nuclei behave. Recall that the nucleus is a very tightly bound collection of protons and neutrons that sits at the centre of every atom, no doubt admiring the dazzling quantum electrons that are orbiting round it. We also know that the nucleus is made of particles with a positive charge (protons) and those with no net electric charge (neutrons). Given that particles with the same type of charge repel each other (see Chapter 3), it must therefore be the case that there is another force, non-electromagnetic in origin, that acts to stick the nucleus together. Furthermore, it must be stronger than the electromagnetic force, as it must counteract the fact that the latter wants the protons to fly apart, rather than being held tightly together in a well-defined bunch. This force is called the *strong nuclear force*, or just the *strong force* for short. Its properties have been well measured since at least the 1960s, in successive generations of particle accelerator experiments.

The fourth and final fundamental force also concerns atomic nuclei. If we watch certain nuclei for a good while, we will occasionally see that a given nucleus suddenly changes into a different type of nucleus. Given that the different types of nuclei (i.e. which chemical element they correspond to) are characterised by how many protons they have, it must be the case that the number of protons has suddenly changed. This can happen in one of two ways: either a proton suddenly turns into a neutron, or vice versa. In the former case, two particles are produced alongside the neutron: a positron and a neutrino. In the opposite case, a neutron turns into a proton, accompanied by an electron and an antineutrino. It was in fact by studying such processes that the neutrino was first discovered. If one measures the total energy carried by the initial and final particles without accounting for the (anti)neutrinos, it is found that the total amount of energy in the universe changes. However, physicists at the time believed so strongly in conservation of energy – for excellent reasons that are still around today – that they stipulated that a new particle must exist that carries away the missing energy. The neutrinos were discovered shortly afterwards.

Both of the above processes are known as *beta decay* and are clearly similar to each other. However, they cannot be electromagnetic in origin: the neutron has no net electric charge and yet still undergoes a decay process, which must therefore be due to some different type of force. Calculations tell us that gravity is much too weak to be causing this effect, and the strong force is much too strong. The relevant force is thus called the *weak nuclear force*, or simply the *weak force*, so as to distinguish it from the strong force. A working theory of the weak force was first written down in the 1970s, earning a Nobel Prize in 1979 for Sheldon Lee Glashow, Abdus Salam, and Steven Weinberg. It has been confirmed in every experiment performed since.

For electromagnetism, we saw that particles carry a property called electric charge, such that only charged particles experience the electromagnetic force. A similar idea is true for all other forces in nature. Gravity, for example, is felt by all objects that have a mass. For the two nuclear forces, we can identify a corresponding charge of fundamental matter particles, whose possession means that the relevant force will be experienced. For the weak force, this is called (*weak*) *hypercharge*, and we will not need to dwell on it too much here. The case of the strong force is more interesting for our purposes. It is felt by the fundamental particles (quarks) that live inside protons, neutrons, and related composite particles. The quarks have a type of charge called *colour*, whose name is purely historical in origin and has nothing to do with actual colours (n.b. the different colours observed by our eyes are an electromagnetic effect and have nothing to do with the strong force!). Whereas electromagnetic charge has two types (+ and −), colour charge comes in three different types, which are conventionally called *red*, *green*, and *blue*. Again, these are just labels, and any other labels would have done. One way to think about colour is to recall that there is a quark field in nature, whose quanta give rise to the quark particles, using the general scheme for QFT that is exemplified in Figures 6.1 and 6.2. At each point in space, we can think of drawing a set of axes representing the amount of red, green, and blue charge that the quark field has at that point. An arrow in this abstract "colour space" corresponds uniquely to a particular set of colour charges, and how this arrow changes as we move throughout space or time tells us how the colour of the quark field is changing.

Our general scheme for QFT tells us that each force in nature must have a corresponding field, whose equations give rise to wave-like solutions. Quanta of these are seen as particles. For the weak force, there actually turn out to be two fields, which give rise to particles known as the *W and Z bosons*. These were discovered in the 1980s at CERN, marking a spectacular success for the theory of weak interactions. For the strong force, the relevant particle is known as the *gluon*, for the simple reason that it glues the proton together! In some later chapters, we often have reason to talk about the theory of the strong force, but where the quarks have been taken out. In other words, we consider the gluons by themselves. This particular toy theory is usually referred to as *Yang–Mills theory*, after the people who discovered it in the 1950s. Interestingly, it was way ahead of its time, and it was only realised that the theory could be used to properly describe the strong nuclear force in the 1970s. For completeness, when the quarks are included as well, the theory has the lovely name of *Quantum Chromodynamics*, or QCD for short.

6.4 The Standard Model of Particle Physics

We have now seen almost all of the key ideas we need to be able to describe the particular QFT that apparently underlies our existence. It is called the *Standard Model of Particle Physics*, given that the term *Particle Physics* describes the study of subatomic objects, such as the fundamental particles in nature. Physicists working in this area usually just refer to the theory as the *Standard Model* for short, although some care is needed as there are other "Standard Models" in other branches of physics (e.g. cosmology).

The simplest way to view the Standard Model is that it is the QFT that describes all of the fundamental matter particles in nature, plus how they interact with three of the four fundamental forces: electromagnetism and the weak/strong nuclear forces. The missing force is gravity, and we see why it is missing in the following chapter. Leaving gravity aside though, the Standard Model is itself a spectacular achievement that distils literally thousands of years of human thought in a single place. Almost all the physics we know about – Quantum Mechanics, Special Relativity, and the laws underlying our everyday lives – emerge from the Standard Model. Yet the equations defining the theory can be written so compactly that they fit on a coffee mug. Indeed, such mugs can be bought, at the time of writing, from the CERN gift shop!

It is clearly beyond the scope of this book to describe the equations of the Standard Model in detail. For a trainee physicist, this would typically happen either in the fourth year of their degree or in the first year of a PhD. Instead, the usual way of describing the theory to non-experts is to simply list the particles that the theory contains, and we have already referred to most of them already. In Table 6.2, I list the various matter particles but in a way that displays more structure than I have hitherto alluded to. The left-hand part of the table lists the various quarks. There are six different types of quarks that have the weird historical names of up (u), down (d), charm (c), strange (s), top (t), and bottom (b), where the latter leads to many an amusingly puerile joke at conferences. For each particle, I have given the mass in certain weird units that particle physicists use. They talk about things called "Giga electron volts" (GeV), such that 1 GeV is equivalent to 1.78 billion-billion-billionths of a kilogram. If you don't understand these units, don't worry, but the numbers for the masses are useful in showing us how the masses of the various matter particles compare with each other. We can see, for example, that the up, down, and strange quarks are relatively light, the charm and bottom quarks are heavier, and the top quark is extremely heavy, indeed the heaviest fundamental particle known in nature.

The right-hand side of the table shows particles called *leptons*, which is a collective term given to particles like the electron (e.g. the muon and tauon) and the various neutrinos, where the latter are denoted by the Greek letter nu (ν). It is common to use short-hand

Table 6.2.   The fundamental matter particles. Shown are the three generations of particles, their masses, and which forces they feel.

		Quarks					Leptons			
		Mass/GeV	Strong	Weak	EM		Mass/GeV	Strong	Weak	EM
1st Generation	$\begin{pmatrix} u \\ d \end{pmatrix}$	0.002 0.005	✓ ✓	✓ ✓	✓ ✓	$\begin{pmatrix} e^- \\ \nu_e \end{pmatrix}$	0.0005 ?	× ×	✓ ✓	✓ ×
2nd Generation	$\begin{pmatrix} c \\ s \end{pmatrix}$	1.28 0.095	✓ ✓	✓ ✓	✓ ✓	$\begin{pmatrix} \mu^- \\ \nu_\mu \end{pmatrix}$	0.106 ?	× ×	✓ ✓	✓ ×
3rd Generation	$\begin{pmatrix} t \\ b \end{pmatrix}$	173 4.18	✓ ✓	✓ ✓	✓ ✓	$\begin{pmatrix} \tau^- \\ \nu_\tau \end{pmatrix}$	1.78 ?	× ×	✓ ✓	✓ ×

symbols for the first of these types of particles, such that the electron, muon, and tauon are denoted by e^-, μ^-, and τ^-, respectively. Again, all of the entries in the table constitute different types of matter particles so that, e.g. there are three entirely different types of neutrinos. We see a large variation in the masses of the electron, muon, and tauon, which are otherwise very similar to each other. We do not yet know the masses of the neutrinos, but we have known definitively since 2001 that neutrinos do indeed have a mass.

In Table 6.2, I have indicated which fundamental forces are experienced by each of the matter particles. We see that the quarks experience strong, weak, and electromagnetic forces. They must feel the strong force, given that they make up the particles (protons and neutrons) inside atomic nuclei, which we know are held together by the strong force. That quarks experience the weak force explains why protons can turn into neutrons and vice versa, and hence why certain atomic nuclei can change into others. Finally, the fact that quarks experience electromagnetic forces tells us that they carry electric charge, which in turn explains why the proton can be electrically charged. For the leptons, we see that they carry the weak force, hence why they can be emitted when atomic nuclei decay. The electron, muon, and tauon experience electromagnetic forces, given that they carry electric charge. The neutrinos are electrically neutral (indeed, their name is Italian for "little neutral one") and thus do not experience electromagnetic forces. Finally, none of the leptons experience the strong force, but we could have worked this out already: if electrons felt the strong force, they would be bound inside atomic nuclei, rather than orbiting outside them as is needed to get atoms to work properly.

Table 6.3.   Fundamental forces and the particles that carry them.

Force	Carrier(s)	Mass/GeV
Electromagnetism	γ (photon)	0
Strong	g (gluon)	0
Weak	$W^{\pm}, Z^0$	80.39 $(W^{\pm})$, 91.188 (Z^0)

The earlier discussion will no doubt come across as particularly dry, given that it is nothing more than a list of particles and their properties. It is easy to lose sight of what Table 6.2 is useful for, and the answer of course is *everything!!!* All of what we see around us – and other stuff that we don't directly see – is made of these particles. If they didn't have the right properties, we wouldn't be able to exist nor would our Sun be able to heat our planet so that we could have evolved to live on it! There are also other potentially interesting things we can notice about Table 6.2. We see that is not just a random shopping list of different types of particles but that there seems to be some sort of regular structure or pattern. That is, I have chosen to group the particles into pairs, such that the three rows of the table look almost identical, up to the fact that the particles in each row have different masses. In any given row, each particle has, apart from its mass, identical properties to its counterpart in any other row of the table, and thus it looks as if nature has decided to give us three copies of the same basic set of particles. Physicists call these *generations*, and it is not completely clear why we have them, although we revisit this point in the following.

Each of the matter particles in Table 6.2 has a corresponding antiparticle. To this, we also need to add the fundamental forces. QFT tells us that these will also be carried by particles, and we show the particles that carry the various forces in Table 6.3. The photon and gluon are denoted by γ and g according to convention. There are then three particles that carry the weak force. Two of these – denoted by W^+ and W^- – have the same mass, which is related to the fact that the W^+ particle turns out to be the antiparticle of the W^-. The third particle, Z^0, has a slightly different mass, and the fact that these masses are not zero has been known since the weak force carriers were discovered at CERN in the 1980s.

The equations of the Standard Model tell us precisely how all of the particles in Tables 6.2 and 6.3 arise and how they mutually interact. The route to the final theory was tortuous, requiring both abstract theoretical reasoning, and detailed experiments to tell us how things should be. We have already seen many examples of how following mathematics rather than everyday intuition was the right thing to do at certain crucial moments in physics history, such as the development of Special Relativity and Quantum Mechanics. At other times, experiment had to guide our species as to the right way forward, which was particularly the case for our description of the weak force.

6.5   Gauge Invariance and the Higgs Boson

There is one particularly famous fundamental particle that you may have heard of, but which is missing from Tables 6.2 and 6.3. It is called the *Higgs boson* and arises for the following reason. The equations of the Standard Model are not completely arbitrary but have a very special structure. Recall that the quark fields that feel the strong force can be thought of as having an abstract "colour space" at each point in space, as shown in Figure 6.3. An arrow in this space tells us how much red, blue, or green charge there is at any given location,

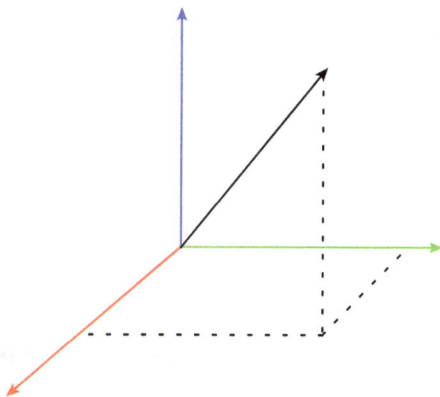

Fig. 6.3.   Quarks carry a type of charge called *colour*, which can come in three types (red, green, and blue). Mathematically, we can think of the quark field as having an abstract *colour space* at each point in space, where an arrow in this space tells us the amount of redness, greenness, and blueness.

such that we can literally think of the quark field as carrying such an arrow as we travel from point to point in space or at different times. Of course, what we happen to call red, green, or blue charge is up to us. We are free to redefine these labels, which in terms of Figure 6.3 means that we can redefine the axes or alternatively rotate the black arrow so that we are measuring colour charge according to a different convention. Doing this is a human choice that cannot have any impact on what happens in nature. Thus, the equations describing the theory must be insensitive to our changing the colour arrows in this way. Physicists have fancy words for this. Whenever equations are insensitive to a particular operation (such as rotating weird arrows in abstract colour spaces), we say that they are *invariant*, meaning "do not change". The particular property alluded to earlier is known as *gauge invariance*, where the word gauge implies "changing how we measure things" (in this case, colour charge). If we change our convention for how to measure colour charge, we must do this in the same way at all points in space so that we are consistent. This is called *global gauge invariance*, where the word "global" here implies "everywhere at once".

Remarkably, all of the field theories in the Standard Model have a much more compelling structure than just this. It turns out that we can choose to rotate the colour arrows such as that shown in Figure 6.3 by different amounts at different points in space and that the mathematics of being able to do this has a remarkable consequence: demanding that the equations describing the quark field have this property *automatically* leads to the existence of the gluon field, which has exactly the right properties to describe the strong force! This is called *local gauge invariance* to distinguish it from its global counterpart, and similar ideas apply to the electromagnetic and weak forces. Historically, local gauge invariance – the ability to redefine what we mean by charge at every individual point in space and/or time – was a guiding principle, without which it would have been impossible to arrive at the final equations of the Standard Model. To some extent, we can claim that local gauge invariance even gives us an explanation of where the different forces come from. However, it has merely shifted the question, as we now have to explain why local gauge invariance should be there in the first place.

A particularly interesting consequence of local gauge invariance – at least the type that is present in the Standard Model – is that it implies that all the fundamental matter particles and force carriers should have no mass, which is clearly at odds with what is observed. There is, however, a way around this, as first shown by three independent research groups in 1964.[1] One can introduce an extra field that the other particles interact with. In doing so, they magically gain a mass, allowing the Standard Model to respect both local gauge invariance, and the fact that many of its particles are massive. As we know by now, wherever there is a field, there are particles. The particle associated with the extra field is the *Higgs boson* mentioned earlier and was first observed in 2012 at CERN.

6.6   Problems in the Standard Model

The Standard Model (SM) of Particle Physics is genuinely one of the crowning achievements of our species. Without the physical understanding it distills, we wouldn't have had any of the technological development that has enhanced our quality of life over the last few hundred years. There remains the ugly and obvious truth, however, that the Standard Model is certainly wrong or, perhaps more diplomatically, incomplete. Its deficiencies include the following:

- *Gravity is missing*: The SM does not include gravity, for the simple reason that we do not know how to turn our best theory of gravity (General Relativity) into a fully quantum theory. We explore this in more detail in the following chapter.
- *Unexplained structure*: Glancing at Table 6.2 reveals a huge range of different masses for the fundamental particles. Are these masses arbitrary numbers that nature has given us? We might hope instead that there is some explanation for where these come from. Likewise, why do we see particular forces (strong, weak, and electromagnetic) and not others?
- *Matter dominates over antimatter*: Our universe is apparently mostly made of matter and not antimatter. If this were not the

---

[1]The three sets of authors were as follows: Brout and Englert, Higgs, and Guralnik, Hagen, and Kibble. The reason for this pedantic footnote is that tensions run rather high in the particle physics community about who exactly did what and when ...!

case, we would see it very clearly: e.g. galaxies made of antimatter would collide with galaxies made of matter, leading to spectacular bursts of energy that would be easily detectable on Earth. But we saw earlier that matter and antimatter must always be created in equal amounts, which leads to the obvious puzzle of where all the antimatter has gone. If one looks more carefully, it turns out that the Standard Model has some of the right properties to explain this, and the fact that there are three generations of particles turns out to be useful. But the SM itself does not fully solve the matter/antimatter problem, such that we believe there must another theory beyond it.

• *Dark matter and dark energy*: Measurements from astrophysics (looking at the sky) suggest that there is a lot of stuff in the universe that is very heavy but which only interacts very weakly with the particles in the Standard Model. This is usually referred to as *dark matter*, an idea which is still somewhat controversial. If it exists, however, we are confident that it must involve some extension to the SM. Similarly, measurements of how the universe is expanding tell us that there is a mysterious substance – *dark energy* – that is causing the universe to *accelerate* in its expansion. This is also not explained at all by the Standard Model.

All of these problems are difficult to argue with, and many candidate theories have been proposed over the years that involve adding extra particles and/or structures to the Standard Model, in order to solve things. Broadly speaking, we know that any new theory of physics must look like a QFT at sufficiently low energies, in the same way that QFT itself reduces to other theories of physics where necessary, as indicated in Table 6.1. This helps in knowing how to write down possible corrections to the SM. However, any new theory that is proposed has to also agree with all previous measurements that have been carried out, which becomes an increasingly complex task as more measurements are made.

Our main way of testing theories that go beyond the Standard Model is to use particle accelerators: large experiments that accelerate beams of particles (e.g. protons, antiprotons, or electrons), before focusing them so that they collide. Large detectors surrounding the collision point capture all the resulting subatomic debris, and by comparing this mess with the results of calculations, we can tell the difference (in principle) between what the detector measures and what we expect from the Standard Model. Any deviation means that we have discovered new physics, which would then need further effort to pin down. At the time of writing, the current flagship particle accelerator is the Large Hadron Collider at CERN, near Geneva. This collides beams of protons that are accelerated by passing them multiple times around a 27 km circular tunnel. This large size is to minimise the energy that will be radiated by the orbiting protons, which in turn imposes a limit on how fast we can make them go. The protons are made to collide at several places around the ring, where the detectors are. Of the thousands of scientists that work at CERN, some will be permanently based there, and others will be based in universities around the world, analysing data remotely and travelling to CERN itself for weeks at a time. Other scientists, like myself, will travel for research conferences or to meet research collaborators. It is a thriving hub of scientific activity, with never a boring conversation over lunch, and even social clubs for more long-term members or visitors. In my own career, I have seen CERN's fame increase as various discoveries have been made and its resulting status as an interesting tourist destination. Nowadays, one can even take a public tour, although access to the tunnels is only possible when the experiment is not running, due to the large amounts of radiation one would otherwise encounter. The vast majority of the machine consists of an apparently unremarkable metal drainpipe, around which an enormous amount of complicated machinery controls the conditions inside: see Figure 6.4.

Quite how large-scale particle physics experiments will evolve in the coming years remains somewhat uncertain. The Large Hadron Collider was explicitly designed to either find the Higgs boson or whatever would replace it should it not have existed. It has performed this task excellently but has so far not provided very strong hints about where to look for corrections to the Standard Model. The next few years are likely to see the development of new particle accelerator technologies, as well as precise studies of Standard Model particles, in order to look for deviations from their expected behaviour.

Fig. 6.4. The beam pipe of the Large Hadron Collider gives a great glimpse of how down-to-earth modern science can be while at the same time being incredibly complicated!
Source: Photo by Erwan Martin on Unsplash.

Summary

In this chapter, we have explored our most complete theory of the fundamental matter particles in nature, as well as most of the forces by which they interact. The key ideas we need for later are as follows:

- QFT combines Special Relativity and Quantum Mechanics in a single consistent framework.
- It tells us that all particles in nature – associated with either matter or forces – arise as quanta of fields, generalising the idea of photons as quanta of the electromagnetic field.

- The theory predicts the existence of antimatter that annihilates matter to give a burst of energy.
- There are four fundamental forces in nature: electromagnetism, the weak and strong nuclear forces, and gravity.
- All forces apart from gravity are described by the Standard Model of Particle Physics, which also describes the fundamental matter particles.
- The strong force is carried by gluons, interacting with quarks. The theory of the strong force by itself, excluding the quarks, is called *Yang–Mills theory*. When the quarks are included, it is called *Quantum Chromodynamics* (QCD).

Chapter 7

A Weighty Subject

In this chapter, we finally address the force of gravity which, as we know from previous chapters, is one of the four fundamental forces in nature. Its conspicuous absence from the Standard Model of Particle Physics demands an explanation, and I start by briefly reviewing what we thought gravity looked like before 1915.

7.1 Gravity before 1915

In Western culture, gravity is strongly associated with Isaac Newton. Since childhood, many of us have been fed strong images of a lone man in a horsehair wig, whose nap under a tree is suddenly disturbed by a large apple falling on his head. Rather than annoying the reclining scientist, he then immediately exclaims his full understanding of the mysteries of gravity, namely the force that pulls us ineluctably towards the ground. Whether or not these events actually happened, there is a lot more to gravity than this, and Newton's theory was an astonishing achievement when it was first developed in the 1600s. To explain his theory from a somewhat more modern point of view, we can use the same terminology that we used to describe electric forces in Chapter 3.

Whereas objects experiencing electric forces have a *charge*, this is only one of the properties that objects can have. As we saw in Chapter 2, they can also have a *mass*. Newton's theory tells us that masses create a gravitational field in a similar way to how electric charges create electric fields, but there is one crucial difference.

We saw that electric charges come in two types, which we convention-ally call positive and negative. Opposite charges attract each other, whereas like charges repel each another. In gravity, there is only one type of "charge", in that mass must always be positive. There is thus only one behaviour for the gravitational force, which is observed to be always attractive. Other than this, the interpretation of the grav-itational field is much like the electric field. It has both a size and a direction at each point, where the size gives us the strength of the gravitational force, and the direction tells us which direction the force points. In other words, if we place a mass in a given gravitational field, it will follow the direction of the gravitational field arrows. As an example, Figure 7.1 shows the gravitational field generated by a mass at the centre of the diagram. We can see that all arrows point towards the mass. Thus, were we to put another stationary mass *anywhere* in the diagram, it will always start to move towards the mass at the centre, which is indeed an attractive gravitational force. You may further like to compare Figure 7.1 with Figure 3.3(b), which shows the electric field of a negative electric charge. The results look almost identical, suggesting a strong analogy between electric forces and Newton's description of gravity. Indeed, an equation can be writ-ten which describes precisely how the gravitational field is generated by masses, and it looks rather like a similar equation for electric fields that forms part of Maxwell's theory.

Figure 7.1 allows us to understand the most commonly encoun-tered feature of gravity, namely that it holds objects on the surface of the Earth, such that they don't float around the room. It turns out that, for large objects, we can pretend for many purposes that all of their mass is concentrated at a certain point within the object, called the *centre of mass*. For ball-shaped objects such as the Earth, this will be at the centre, and thus Figure 7.1 tells us that gravity is trying to pull all nearby objects in to the centre of the Earth. What stops this from happening are other forces (e.g. electromagnetism and certain quantum effects) that refuse to let us and our belongings pass through the floor. Were these other forces not there, we would indeed carry on through the Earth's crust and eventually end up bobbing about at the Earth's core, where we would find it rather warm at any time of the year. Note that, far from the centre of the Earth (e.g. towards the top of Figure 7.1), it looks roughly like all of the arrows are pointing in the *same direction*, if we do not go too far to the left or

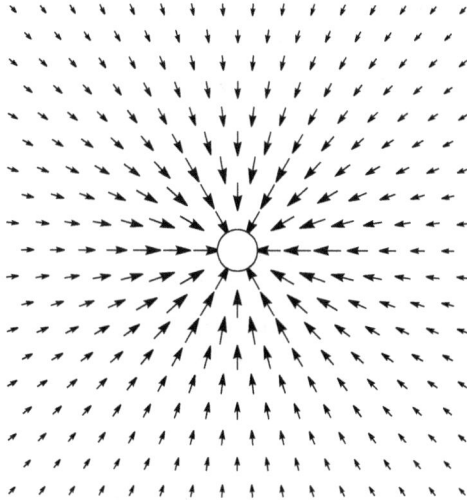

Fig. 7.1. In Newton's theory of gravity, masses generate a gravitational field, which can be represented as an arrow at each point in space.

right. This explains why, at any point on the Earth's surface, it looks as if gravity is pulling everything downwards. However, this is only a close approximation to what is actually happening, which is that the gravitational field points in slightly different directions at different points but *always* towards the Earth's centre of mass!

Newton's theory of gravity was revolutionary at the time. An immediate consequence of it, for example, is that masses anywhere in the universe obey the same force of gravity. By combining his equations for gravity with the laws of motion, one can not only predict the motion of objects on the Earth – from footballs to missiles – but also the orbits of the planets. Before Newton's theory, massive amounts of very precise data had been collected about how the planets move, and it was found that their orbits always obeyed certain conditions. Newton's theory explained all of these conditions at a stroke and also firmly established the idea – in Western culture at least – that the motion of "heavenly bodies" was ultimately no different to motion in our own immediate world. So completely do we take this idea for granted nowadays that we have trouble even understanding how people would have had a problem with it. The power of Newton's theory also led, in subsequent centuries, to new astronomical discoveries. In the 1800s, for example, the French astronomer

Urbain Jean Joseph Le Verrier found that the orbit of Uranus was not behaving properly, according to his prediction from Newton's law of gravity. He could patch up the calculations if he assumed that there was a new planet, whose position could be precisely calculated. He sent his prediction to Johann Gottfried Galle at the Berlin observatory, who found Neptune on the same night as he received Le Verrier's letter.[1]

Despite its far-reaching power, Newton's theory of gravity is not without its problems, which are similar to those we already encountered for electromagnetism. In particular, Newton's theory involves the same peculiar property of "action at a distance" that troubles naïve theories of the electric force. If we have a mass on one side of the universe and a mass at the other end, Newton's equations tell us that the force of gravity on the second mass will instantly change if we move the first mass. If this seemed unlikely in Newton's time, it must certainly be regarded as impossible once we know about Special Relativity, which tells us that no information can travel faster than the speed of light. This led Einstein and others to seek an alternative theory that somehow reconciles gravity with Special Relativity, and the result became known as the *General Theory of Relativity*, or just General Relativity (GR) for short. It was first published in 1915, and the name is a somewhat unfortunate one, in that it does not actually tell us that this is first and foremost a theory of gravity. However, what it tells us about where gravity actually comes from is so delightfully bonkers that no name could possibly do the theory any justice.

7.2   General Relativity

The key observation that started Einstein and others on the path to General Relativity is a fact about mass that we overlooked earlier. We first described mass in Chapter 2, where we saw that it measures how resistant objects are to having their motion changed. That is, mass represents the intrinsic *inertia* of an object and in this context

---

[1]An independent set of calculations was made earlier than Le Verrier by the great Cornish astronomer John Couch Adams but did not lead to such an immediate experimental search.

is sometimes formally called *inertial mass*. In this chapter, we have seen mass playing a different role, as the analogue of electric charge that generates gravitational fields, rather than electric fields. In this context, we would call this *gravitational mass*, and the reason for drawing such a pedantic distinction between these two concepts of mass is that it is not obvious beforehand that these should be the same thing. Inertial mass can be measured by looking at the resistance an object displays upon responding to any force, not necessarily a gravitational one. If gravity is not at all involved in making such a measurement, why would we assume that this inertial mass is the same thing that generates gravitational fields?

The curious thing about the universe we live in is that inertial and gravitational mass do indeed turn out to be the same thing, as has been verified by careful experiments. This fact is usually referred to as the *equivalence principle*, where this phrase is used so often by physicists working on gravity that they know in every case that it is inertial and gravitational masses that are being talked about as being equivalent. What will not at all be obvious is quite how profound the consequences are of what, on the face of it, is a very simple statement. To understand this in more detail, let us interpret the equivalence principle in a different way. Inertial mass describes how an object's trajectory changes in response to a given force, whereas gravitational mass dictates how an object experiences a gravitational field. What the equivalence principle then tells us is that it is completely impossible to tell apart the effects of an object experiencing a gravitational field, from it simply accelerating along some trajectory, *as if* there was a gravitational field. Thus, we can pretend that there is no gravitational field, provided that we make sure that objects have trajectories that look the same as if there was one.

Now imagine that there are no forces acting on a given object other than the gravitational force. Objects subject to no forces usually move at constant speed, which amounts to moving in a straight line through space. Now, however, we are requiring that an object with only gravity acting on it must move along a non-straight line, namely some curved path, so that it looks like there is a gravitational field. If we really want to pretend that there is no gravitational force, there is only one way to do this: we can say that *space itself* has become curved so that objects no longer move along straight lines but are forced to follow curved ones. Provided we arrange this "curved" space in just the right way, objects moving in it will look like they experience gravity. Put more simply: we can think of the gravitational force as arising from replacing our notion of space and time with a *curved spacetime*. Objects moving in curved spacetime follow curvy trajectories, which can move towards each other to mimic the force of gravity.

This is all very heavy stuff and required a huge leap of imagination from Einstein and others, who also spent about a decade learning the relevant mathematics to describe curved spaces. But there is a simple analogy that may help. Imagine that you have two pet ants and that you place them on a large sphere, such as a beach ball. I have drawn this ball in Figure 7.2, and we can imagine that we place the ants at the positions marked A and B in the figure. Now imagine that your ants are sufficiently well trained that they can follow your every command, and consider what happens if you tell them to walk towards the "north pole" of the ball, marked N. As far as each ant is concerned, there are no obvious forces acting on them. They will assume this given that, if the sphere is large, they will think they are simply moving in a straight line, similar to how, in your everyday life, you are fooled into thinking that you can walk in straight lines, when you are really following the curved surface of the Earth. As the ants make their way towards the point N, they will get gradually closer together, and thus it looks as if there is

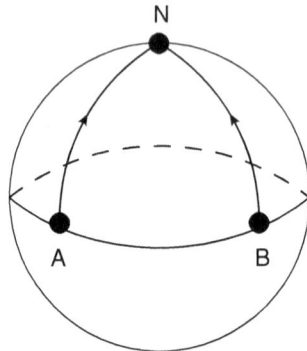

Fig. 7.2.   If two ants start walking towards the north pole N on this ball, starting from points A and B, it will look as if an attractive force has brought them together. In fact, there is no such force, and it is the curvature of the surface that has made them meet.

an attractive force between them. We know, however, that there is no such thing: it is simply that the space they are moving on is curved!

I find this a very helpful analogy for how curved spaces can mimic attractive forces. It is, however, a flawed analogy. In particular, you can think of constructing a surface (e.g. the top of a vase) that curves *outwards* rather than inwards. Then, the curved surface would produce a repulsive force rather than an attractive force. This does not happen in Einstein's theory of gravity: there is only ever an attractive force. Furthermore, in Figure 7.2, we have only considered the fact that space is curved, but in Einstein's theory, we must end up combining our three dimensions of space with the time dimension, to make a four-dimensional *spacetime*. It is this four-dimensional thing that is curved, which is utterly impossible to visualise. However, the mathematics of how to describe curved spaces – in any number of dimensions – is very well formulated and tells us exactly how to go about things.

What, though, does it even mean to say that spacetime is curved? And what is it that causes this curvature? In General Relativity, it is the presence of mass (or, equivalently, energy) at a given location in spacetime that causes it to curve. The more mass or energy there is, the more curvature there is, and thus the stronger the apparent force of gravity, as expected. A well-worn analogy is again helpful, where it helps to think of space as two-dimensional rather than three. Our normal flat space can then be thought of as a constant fixed sheet, upon which objects can move about. The spacetime of General Relativity, however, is more like a sheet made of rubber. If you place heavy objects on such a sheet, they will deform it as shown in Figure 7.3. After that, any smaller objects such as marbles will no longer follow straight lines on the sheet but will instead follow curved paths around the heavier object. This is essentially what is happening when planets orbit the Sun, according to General Relativity: the Sun causes a large curvature of spacetime. Each planet then tries to follow a straight line, but this gets warped due to the underlying curvature of the spacetime so that the various orbital paths are followed instead.

As objects move, their location in spacetime changes. This tells us that the curvature of spacetime is itself continually changing and evolving. We might think about the difference between this and our

Fig. 7.3.   Heavy objects curve spacetime, causing smaller objects to orbit around them.
Source: Photo by Hassaan Qaiser on Unsplash.

traditional notions of space and time by considering actors performing in a normal theatre, with a fixed wooden stage. If we replaced
the stage with the surface of a trampoline, we would be heavily dis-
tracted during the performance – as would the actors – by the fact
that the stage kept wobbling about in response to how they moved.
If we take Shakespeare literally when he tells us that "all the world's
a stage", then we are not far off understanding General Relativity!

It is now hopefully clear what General Relativity has to do. In
order to be a useful theory of physics, it must give us a precise
set of equations that tell us, for an arbitrary collection of mass
and/or energy scattered throughout the universe, what the structure of spacetime looks like. These equations do indeed exist and
are usually referred to as the *Einstein equations*. Of course, Einstein
wrote a great many equations during his lifetime. It tells us quite
how revered General Relativity is, that these equations are somehow
taken as his greatest achievement. They are also some of the most
difficult equations to solve in the whole of physics and are usually
encountered in the third or fourth year of a physics degree. However,
it is worth mentioning one necessary feature of them. In earlier chapters, we became used to the fact that what we think of as perfectly
sensible physical theories often end up being replaced by some seriously weird alternative, whose mathematics tells us that our everyday
notions of familiar things are not to be trusted. General Relativity
is yet another example of this, but we know that in many situations,
Newton's original theory of gravity works perfectly well. It must be

the case, therefore, that when the gravitational force is sufficiently weak (e.g. far away from the Sun), Einstein's equations reduce to Newton's. This is indeed the case and was one of the guiding principles used by Einstein in finding his equations in the first place.

Having now described the main ideas of General Relativity, let us look at some of its consequences.

7.3 Consequences of General Relativity

7.3.1 *Light bending*

In traditional flat space, light must travel in straight lines. You may have seen this before in everyday life, such as when rays of light from the Sun becomes visible as in Figure 7.4. When gravity is present, General Relativity tells us that we must replace flat space with a curved space, where the amount of curvature is determined by how much mass is present. In simple terms, light rays will then get bent

Fig. 7.4.   In flat space, light travels in straight lines, as exemplified by rays of light from the Sun.

Source: Photo by Jasper Gronewold on Unsplash.

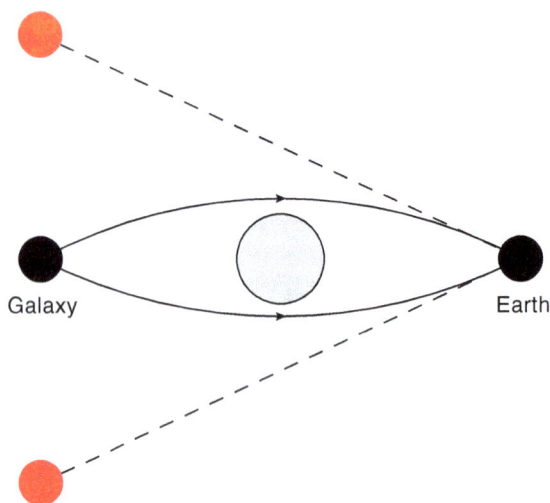

Fig. 7.5.   In *gravitational lensing*, light from a distant galaxy (shown in black on the left) can be bent around a heavy object, such that it reaches the Earth from multiple directions. An observer on the Earth then sees multiple images of the galaxy, as if the light has come from the locations shown in red.

by heavy objects. It takes a seriously heavy object to see this effect, and it was first measured by looking at how the light from distant stars gets bent by our own Sun.

Another novel example of this effect is that of *gravitational lensing*. Imagine that an astronomer wants to observe a distant galaxy, whose light can reach the Earth. If there is a highly massive galaxy in between, the light from the distant galaxy will be bent, such that it reaches the Earth from different directions, as shown in Figure 7.5. The astronomer on the Earth will then see multiple images of the galaxy, as if the light had travelled in straight lines from far away. This effect is in fact incredibly useful. The bending of the light rays means that we can see distant galaxies that we wouldn't be able to see otherwise (i.e. they would be blocked by the massive galaxy in the middle of Figure 7.5). Also, given that the amount of light bending depends on the mass of the middle galaxy, we can use gravitational lensing measurements to get estimates of the amount of matter in certain galaxies. Where this greatly exceeds the amount of matter that we can actually see, it provides evidence for the dark matter that was mentioned in the previous chapter.

Far from being an abstract concept that is only of use to professional astronomers, the bending of light – and its necessary implication that the spacetime we live in is curved – is nowadays part of our everyday lives. When you use a map on your smartphone, you are able to pinpoint your location to within a few metres. The computations being carried out in the phone, which relate to how the signals it sends are being bounced off satellites in orbit around the Earth, have to take into account the curvature of spacetime in order to get the needed precision. Thus, General Relativity is much more practically useful than it might first appear!

7.3.2 Black holes

If light gets bent by travelling through curved spacetime, it begs the question of whether there are regions of spacetime that become so warped, that light cannot escape from them. General Relativity indeed predicts such behaviour, meaning that there are various solutions of Einstein's equations in which some part of spacetime is "cut off" from whatever lies outside it. Such objects are called *black holes* and were one of the first consequences to be derived from Einstein's equations. As we will see, we can talk about black holes as having a certain size, and we can also talk about material that may be inside a black hole. However, no communication can take place between the inside of the black hole and the outside. Furthermore, if anything is unfortunate enough to fall into the black hole from outside, then it will remain stuck inside forever.

The simplest black hole is called the *Schwarzschild black hole*, named after the person who discovered it barely a month after Einstein's equations first appeared. The Schwarzschild black hole is particularly special in that the material inside it carries no net electric charge and nor is it spinning or rotating. This means that the "shape" of the black hole in space is perfectly ball-shaped, and we can draw it as in Figure 7.6. There are a number of interesting features. First, there is the fact that the centre of the black hole will have a certain location in space, which is assumed to be stationary when solving the equations of General Relativity. These equations in turn tell us the size of the region from which light cannot escape, corresponding to a certain radius for the ball, as depicted in the figure. It is called the *Schwarzschild radius* and depends on the mass inside the black hole

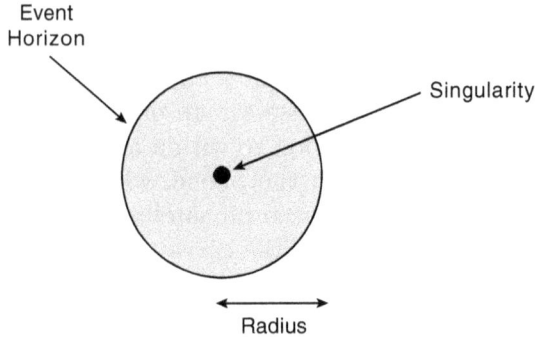

Fig. 7.6.   A Schwarzschild black hole consists of a ball-shaped region of space-time, from which nothing (even light) can escape. At its centre, the curvature of spacetime becomes infinite, and the equations of General Relativity break down.

and the strength of the gravitational force. In particular, the radius increases with mass, such that if additional material falls inside the black hole, the latter will become ever larger. The edge of the black hole is known as the *event horizon* and marks the boundary between things that can escape the black hole and things that cannot.

Historically, the notion of black holes was very difficult for many physicists to accept. One very troubling feature of them, as predicted by General Relativity, is that the curvature of spacetime becomes larger and larger as we move towards the centre of the black hole. At the centre itself, spacetime becomes literally infinitely warped so that the mathematics of General Relativity – which predicted the black hole in the first place – no longer makes any sense. We call this ill-behaved point right at the centre of the black hole a *singularity*, and it is extremely troublesome in our trying to interpret black hole physics. How are we to accept that such objects may exist in nature?

One resolution might be to say that the assumptions that go into deriving Schwarzschild's solution are somehow too special, and never realised in our universe in practice. For example, the most common way of actually making a black hole is if a star stops shining and collapses in on itself. When this happens, it is virtually impossible that this would happen without at least some of its material rotating. Also, this material would have some net electric charge in general, and neither of these factors is accounted for by the Schwarzschild black hole. However, rotating and charged black hole solutions of General Relativity do indeed exist and are well known to physicists. The most general solution – which includes the effects of non-zero mass, rotation, and electric charge of the black hole – is called the *Kerr–Newman black hole*. It has a complicated structure involving multiple event horizons and most certainly has a singularity at the centre. Indeed, certain mathematical theorems tell us that, in the conventional theory of General Relativity, singularities are inevitable and that there is no way to avoid them. An artist's impression of a rotating black hole is shown in Figure 7.7. As it rotates, it collects infalling material to make a hot glowing structure known as an *accretion disk*.

Trying to interpret whether black hole solutions make sense gets even trickier given that we now have a great deal of observational evidence for them. The most compelling recent discoveries include the direct imaging of a rotating black hole by the Event Horizon Telescope Collaboration in 2019. Detailed comparison of their images with theoretical calculations showed spectacular agreement with General Relativity. Further recent evidence comes from gravitational wave experiments (see Chapter 11), which directly measure radiation given off by colliding black holes.

The subject of what really happens inside a black hole remains somewhat speculative and controversial. It is perhaps not too contentious to say that most physicists currently think that, while General Relativity gets overall features of black hole physics right, it

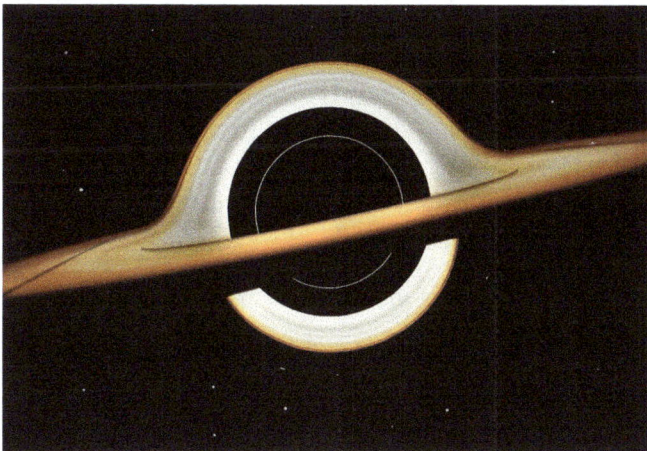

Fig. 7.7.   Artist's impression of a rotating black hole.
Source: Image by Aman Pal on Unsplash.

must be modified somehow to say what happens at the centre, such that the problem of singularities gets resolved.

7.3.3　The Big Bang

The equations of General Relativity in principle tell us what the structure of the entire universe is, given that they predict what the curvature of spacetime looks like for all points in space and at all times. For things like black holes, the curvature of space usually decreases rapidly as we move away from the black hole itself so that we can take these solutions of the equations to represent localised objects. Further away from the black hole we would need to modify the solution to include other objects that may be present.

However, we can be much more ambitious than this and ask what happens if we solve for the structure of the entire universe all at once. Unsurprisingly, different types of solutions are possible, depending on what assumptions we make. A common way of proceeding is to assume that, if we zoom out onto very large scales, the various stars and galaxies that fill our universe all blur into a soupy mixture with no preferred direction. If we make this assumption, we can solve Einstein's equations exactly, and the origin and fate of the universe then depend on precisely how much matter there is in total in the universe. Physicists call such solutions *Friedmann–Lemaître–Robertson–Walker* solutions, and each of them has the remarkable consequence that the universe has not been around forever but started at some definite time in the past, at which it had zero size. This has become known as the *Big Bang theory* and comes rigorously out of solving the equations of General Relativity, subject to the assumptions described earlier. After the initial "explosion" that starts the universe, there are three possibilities for what happens thereafter:

(i) The universe keeps expanding forever.
(ii) The universe expands until it reaches a maximum size and then stops expanding.
(iii) The universe expands until it reaches a maximum size and then shrinks and collapses.

For each of these possibilities, we can draw a graph of the "size" of the universe as time evolves, as shown in Figure 7.8. What may already be obvious is that cases (ii) and (iii) are what happens if

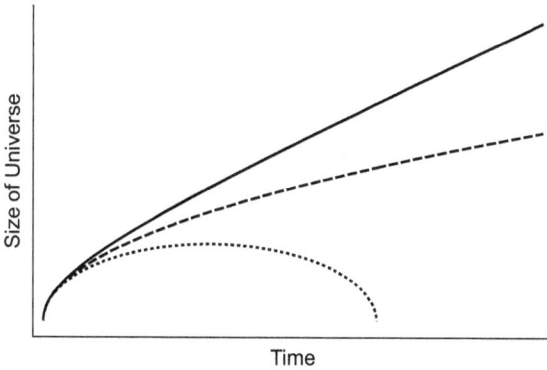

Fig. 7.8. Three possible histories for the size of the universe, against time since the Big Bang: (i) the universe keeps expanding forever (solid); (ii) the universe stops expanding upon reaching a maximum size (dashed); (iii) the universe stops expanding before collapsing back into a point (dotted).

there is *more* matter in the universe: this will have a larger gravitational pull, which will counteract the expansion of the universe. Of course, measurements of the universe we actually live in must tell us which of these possibilities is realised or whether more sophisticated assumptions need to be used. Until relatively recently, it was thought that option (ii) or possibly (i) was what was actually happening, and a success of this approach is that it correctly predicts the relative abundance of hydrogen and helium that we see in outer space.[2] For various reasons, however, it is now thought that there was a period of extremely rapid expansion at the very beginning of the universe, known as *inflation*, whose origin is still not fully understood. Furthermore, recent measurements have shown us that the expansion of the universe is accelerating. This requires the existence of the so-called *dark energy* that was mentioned in the previous chapter and whose origin is still not fully understood.

Even if inflation is included, the presence of a Big Bang itself poses a problem for General Relativity that is similar to what happens at the centre of black holes. If we run time backwards, the universe gets smaller and smaller, until it eventually ends up as a tiny point-like object, whose curvature (warping of spacetime) becomes

---

[2] As the universe expands, it cools, leading to the formation of certain atoms at various times in the history of the universe.

infinitely large. Thus, there is a singularity at the beginning of the universe, and the equations of General Relativity once again break down. The Big Bang is astonishing in that it brings fundamental questions about the nature of our existence within the realms of physical – and directly observable – science. However, answers regarding what happened "before" the Big Bang, or whether such questions even make sense, remain elusive.

7.3.4 *Gravitational waves*

Earlier, we used the analogy of a rubber sheet to understand the warping of spacetime by massive objects. This analogy helps us to understand yet another prediction of General Relativity that is completely absent in Newton's original theory of gravity. Imagine that you and someone else are holding the sheet tight at both ends. Then, consider that your friend wobbles the sheet up and down. What you will see in that case are small wave-like ripples that travel from your friend's end of the sheet to you, whether or not there is anything else on the sheet. Something very similar happens in General Relativity, in that the equations turn out to have wave-like solutions. Then, given that the equations of General Relativity describe the structure of spacetime, it follows that these waves must be ripples *in the fabric of spacetime itself!* They are called *gravitational waves*, and one way of thinking about them is that they are the General Relativity equivalent of the light waves that we saw in electromagnetism. The standard way of deriving gravitational wave solutions is to assume that the gravitational field is weak so that one may write the structure of spacetime as being essentially flat space, plus a small correction. Low and behold, this correction looks like a field filling all of (flat) space, and it is this field that contains disturbances that travel in a wave-like way. This field is called the *graviton field* and can be thought of as the proper gravitational analogue of the electric and magnetic fields in electromagnetism. The equations for the graviton field also reduce to Newton's theory of gravity, whenever the latter is sufficient.

Gravitational waves are a very firm prediction of General Relativity, and the main effect they have is that they cause small deformations in anything they pass through, given that they disturb the structure of space. However, these deformations are so ridiculously tiny that it takes enormous ingenuity to design an experiment that will directly observe them. This was achieved for the first time in 2015 by the LIGO experiment, almost exactly one hundred years after General Relativity first appeared. We see what is involved in such experiments in Chapter 11, and their success in recent years has opened up an entirely new window through which we can view our universe.

7.4   Is Gravity Quantum?

We have seen that General Relativity breaks down at extreme places in the universe, namely the centres of black holes and the beginning of the universe itself. In both of these cases, very short distances are involved, and this should remind us of something. In Chapter 5, we saw that the laws both of motion and of electromagnetism had to be replaced at short distances, by quantum mechanics. Does it not therefore seem sensible that gravity should be quantum too? Indeed, as we saw in the last chapter, the three other fundamental forces of nature do have quantum descriptions. They are collected together in the Standard Model of Particle Physics, which is a quantum field theory. Recall that the basic idea of QFT is that whenever equations describing a field have wave-like solutions, those waves can arrive in well-defined lumps or quanta, which can be thought of as particles.

In this chapter, we have seen that General Relativity itself can be thought of as a field theory, by introducing the graviton field. Furthermore, this has solutions that are wave-like. So what is preventing us from quantising gravity? The application of QFT techniques to describe gravity has been ongoing since the 1960s and is nowhere near as straightforward as the other forces, which are already complicated enough. One can indeed try to apply QFT techniques to the graviton field, which gets its name from the fact that the quanta of gravitational wave solutions are hypothetical particles known as *gravitons*. The problems occur, however, when we try to describe what happens when gravitons interact with each other, as we see in the following chapter.

Summary

In this chapter, we have finally seen how to describe all of the theories that describe our universe, i.e. all of the fundamental forces, in addition to the matter particles. We have also learnt the following:

- Newton's theory of gravity tells us that gravitational forces are produced by massive particles.
- To be consistent with Special Relativity, Newton's theory must be replaced.
- Our modern description of gravity is General Relativity. It says that mass curves spacetime, such that objects moving in this curved space look like they are attracted to each other, hence the force of gravity.
- Consequences of General Relativity include the bending of light by massive objects, black holes, the Big Bang theory of the expanding universe, and gravitational waves.
- General Relativity itself breaks down at the centres of black holes and at the Big Bang itself.
- A quantum theory of gravity may resolve these issues.

Intermission

In the prologue, we saw that a bunch of 30 or so physicists met up in Edinburgh in 2014 and chatted all week in front of blackboards of algebra. I can now tell you the name of this workshop, which was *QCD Meets Gravity*. As we saw in Chapter 6, QCD is the name of the theory that describes the strong force in nature that, among other things, binds protons and neutrons together to make atomic nuclei. Gravity is much more familiar, and the name of the workshop suggests that it was somehow about combining ideas from different branches of physics. Might it turn out, for example, that our very different theories – from the Standard Model on the one hand to General Relativity on the other – are much more closely related than previously thought possible? This does indeed seem to be the case, and it is this idea that had gotten 30 people so excited that they wanted to fly half way around the world. The excitement only increased as the week went by, leading to the agreement to hold the meeting again in 2015.

It didn't stop there. The QCD Meets Gravity workshop was held in Los Angeles in 2015, 2016, and 2017, Stockholm in 2018, and Los Angeles once again in 2019. The year 2020 was much more unusual: a global pandemic halted travel activities around the world, confining people to their homes. However, this didn't seem to stop progress in physics. Instead, scientists – who have long been used to international research collaborations that proceed via email and online meetings – found new and inventive ways to hold their conferences. Talks were delivered using online platforms and recorded to be shared after the event on YouTube. Social Media channels were created so that people could pose and answer questions as the week went on. Online software was used for "virtual" coffee breaks, mimicking the incidental conversations and encounters that have always been one of the most rewarding parts of bringing scientists together. Many of these ideas have stayed around since the pandemic ended, largely due to the democratising effect of enabling many more people to take part in a conference than those who are able to turn up in person. By 2020, the QCD Meets Gravity meeting had grown to nearly 250 participants, from over 23 different countries. The make-up of the conference was also changing: more and more astronomers were turning up and talking with researchers in particle physics and/or pure mathematics whom they wouldn't normally speak to. Something special was happening, and interest in this topic has continued up to the time of writing.

I was lucky enough to be invited to give a talk at the 2020 online meeting, which would have been a highly unusual event even without the pandemic. I had become a father for the first time and was nearing the end of six months of blissful parental leave. My reticence in accepting the talk invitation was quashed by my husband, who suggested I indeed give a talk, to gain the confidence needed to return to work the following week. That this plan worked is largely due to the enormous kindness and friendliness of my fellow scientists, as well as the amazing sleeping ability of little baby Toby. His refusal to inherit his papa's chronic insomnia meant that I had cobbled together enough material to present at the meeting, while giving a remarkable impression of knowing what I was talking about, even managing to not be covered in milk stains (or worse). What was much more interesting, however, was the breadth of topics presented at the meeting, ranging from the pure mathematics of how to do calculations in QCD in the one hand to how to apply this to gravity on the other.

What, however, was the precise nature of the mysterious relationship between our theories of physics that had brought so many people together? To understand this, we need to examine the types of things that scientists who work on these theories actually calculate and why.

Chapter 8

When Particles Collide

In the first half of this book, we have reviewed our best current understanding of how our universe behaves. All things are made out of matter, acted on by only four different types of force. Matter and three of the forces are unified into a single type of theory called a quantum field theory. The fourth force, gravity, is described by general relativity. Although we can choose to interpret this as a field theory by introducing a so-called *graviton field* that gives rise to the force of gravity, the theory itself is not on the same footing as the other forces, as we cannot include the effects of Quantum Mechanics. In this chapter, I want to explore in much more detail why this is the case. To do so, I will need to talk about the kinds of things we calculate in field theories, and why.

8.1 Particle Scattering

As we saw in Chapter 6, our main way of testing the Standard Model of Particle Physics – and the possible theories that may lie beyond it – is to take beams of particles and collide them. When this happens, a variety of other particles may be produced, which are then captured in large detectors surrounding the collision point. Each individual collision of particles is usually called a *scattering event*, and if we want to understand what happens in such events, we need to do some sort of calculation in our theory of interest and then compare this with what is seen in the detectors. We will then get an idea of which of our theories best fits the data. When doing this, a key aspect of QFT is that the Q stands for "Quantum", and Chapter 5 taught us that quantum mechanics means we cannot say what will definitely happen in any situation. Instead, we can only ever say what the range of options is for any given measurement, each of which has some likelihood, or probability, of occurring. To apply this idea to our scattering events, let us assume that we have a given set of particles coming in to the collision point. For this *initial state* of our scattering event, there will then be a number of different options for what happens after the collision, which we call the *final state*. Each of these final states has a given probability of occurring, which we can calculate, at least in principle, from quantum field theory.

What may or may not be obvious in the above discussion is that it makes no sense to measure a single scattering event in the detector and compare this with our given quantum field theory. A single event simply tells us that we correctly predicted one of the possible final states, but it cannot tell us how likely this outcome was to occur. Instead, we must measure *lots of events*, after which we will see that some final states appear more often than others. Provided this pattern of final states matches with the likelihoods we find from the theory, we can confidently say that our theory "matches" the data, at least up to some uncertainty. Where quantum mechanics is involved, this is the best we can do, and it goes almost without saying that the statistical methods involved in contemporary particle physics experiments are extraordinarily complicated, filling entire textbooks by themselves!

To give an example of this procedure in action, let us take a look at Figure 8.1, which includes some data that were measured by the ATLAS experiment at the Large Hadron Collider in CERN. The scientists have tried to isolate all scattering events that contain a top quark and an anti-top quark in the final state, given that these can be produced together. However, (anti-)top quarks do not live for very long before decaying, which can produce an electron and antimuon. The scientists have then measured lots of such events and in each case have asked the following: what is the combined energy of the electron and the antimuon? The way to read the plot is like a bar chart: each bar represents a range of possible values for the energy, all of which are predicted as possible from the equations of the Standard Model. Then, the height of each bar corresponds to how many events are observed within that particular range of values.

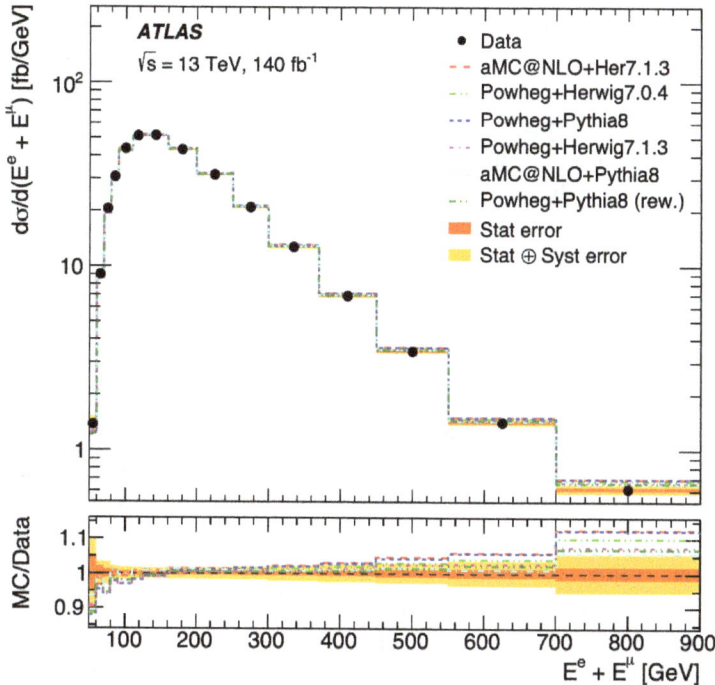

Fig. 8.1. Some actual measured data from the Large Hadron Collider, compared to a range of theoretical predictions from the Standard Model of particle physics.

Source: ATLAS Experiment © 2022.

The dots represent what the experiment measures, and the various coloured lines represent a range of theoretical predictions from the Standard Model of particle physics. In this case, we can see that the Standard Model fits the data very well and thus that there is no evidence yet for any new physics beyond the Standard Model. By compiling hundreds, if not thousands, of similar plots, scientists at the Large Hadron Collider are constantly looking for tiny clues that the SM may be breaking down.

You might be wondering why I have talked about a *range* of possible theoretical predictions. After all, there is only one Standard Model of Particle Physics, so how is it possible that we can get different predictions from the same theory? The answer to this is that it is not usually possible to calculate quantities in quantum field theory exactly, due to the highly complex nature of the equations involved. Instead, approximations are made, where we know, at least in principle, how to improve our approximations by making ever more elaborate calculations. Sometimes, we can compare different types of approximations, to try and figure out which is the best to make in any given situation. This is what is happening in Figure 8.1, although this still does not tell us how the quantum field theory calculations actually work. Let us explore this next.

8.2 Scattering Amplitudes

A typical scattering event consists of a set of incoming particles and a set of outgoing particles, where these will have some particular properties, such as masses, charges, energies, and so on. There are usually only two incoming particles, corresponding to the fact that particle accelerators typically have two colliding beams: getting two beams to collide is already difficult enough, without making things even more complicated by requiring three particles to crash into each other at the same point. If you doubt this, try throwing a tennis ball at a tennis ball that is being thrown by a friend, where your aim is to get the tennis balls to collide in mid-air. Now try this with two friends, and you will see what I mean!

The final state can have any number of particles in general, and we can therefore think of our scattering event as in Figure 8.2. The arrows represent individual particles, and we have drawn time as running from left to right. The event starts out with two incoming particles, and then something interesting happens at intermediate times, when the particles collide. As a result of this interaction, the final state particles emerge and proceed outwards on the right-hand side of the diagram. The central aim of QFT in this context is to somehow work out the probability for observing the given final state on the right, subject to the initial state on the left. And it turns out that there is a single number, called the *scattering amplitude*, associated with any such scattering event. The rules of QFT, at least in principle, tell us how to calculate the scattering amplitude for any choice of initial and final states (that are allowed by the theory being considered). Once we have the scattering amplitude, we can relate it to all of the things that are measured by experiments, such as the data shown in Figure 8.1.

Although the scattering amplitude is a single number, it can change as we vary the properties of the initial and final state particles (e.g. their energies), which makes sense: the probability to have a certain set of energies is not necessarily the same as the probability to have a different set. Thus, our usual way of thinking about scattering amplitudes is to define them for a given set of incoming and outgoing particles, but such that they depend – in a potentially complicated way – on the energies of these particles, and related quantities that characterise their motion. The standard way of comparing QFT to experiments is then as follows: (i) decide which type of scattering event you want to look at (possibly more than one type), (ii) calculate the scattering amplitudes for these events using the rules of QFT, and (iii) relate these scattering amplitudes to quantities measured by the experiment. The calculation of scattering amplitudes has, by now, been going on for many decades but continues to evolve as we seek to make our calculations ever more precise.

8.3 Feynman Diagrams and Rules

While Figure 8.2 is a useful picture, it does not quite contain enough detail. For a start, we could have different kinds of particle, so that we would ideally use different types of symbols for each different

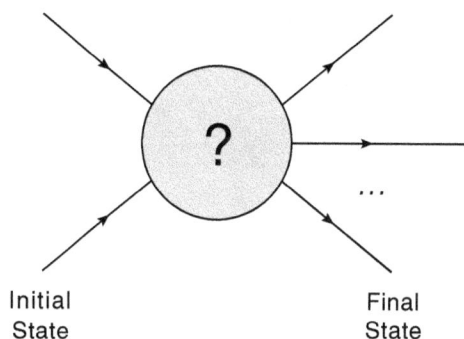

Initial State   Final State

Fig. 8.2.   A scattering event consists of an *initial state* of two particles and a *final state* of any number of particles, where these are represented by arrows, and time runs from left to right. At intermediate times, some sort of interaction occurs, and the job of quantum field theory is to mathematically express the likelihood for this to happen in terms of a number called the *scattering amplitude*.

particle type, rather than denoting everything with a solid arrow. Second, we have not attempted to specify what is going on when the particles interact in the centre of the diagram, and we can in fact be much more precise about this. As an example, let us examine the theory of interacting gluons. Recall that gluons are the particles that bind quarks together to make protons, neutrons, and related particles. We also saw in Chapter 6 that when we consider gluons by themselves, this is referred to as *Yang–Mills theory*. Restricting to a single particle type will allow us to illustrate the main ideas about how we calculate scattering amplitudes, without additional complications. Our central question, then, is what happens when gluons interact among themselves?

Possible diagrams for what happens are shown in Figure 8.3. Again time is running from left to right, and we have chosen an initial state containing two gluons, as well as a final state containing two gluons, although these may have different energies, etc. to the initial gluons. I have also adopted the convention whereby a certain type of curly line represents a gluon. With time running from left to right as before, the way to read Figure 8.3(a) is that two gluons come together and then fuse to make a single gluon, which the rules of QFT certainly allow. Later on, this gluon splits into two individual gluons, leading to the correct final state. Figure 8.3(b) constitutes a different way that we can obtain the same final state: one of the

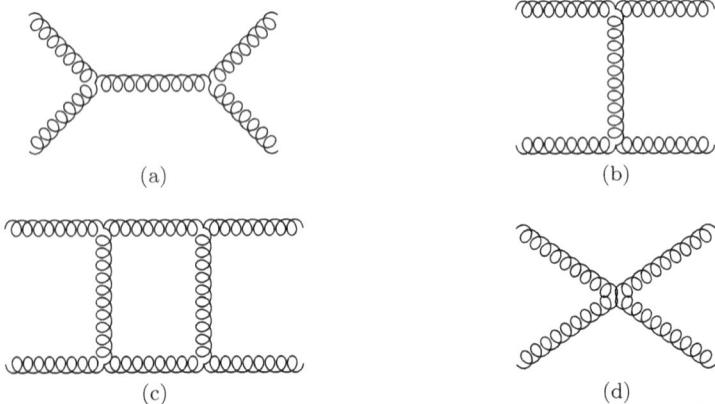

Fig. 8.3.   Different ways in which gluons (shown as curly lines) can interact, producing the same final state. As in Figure 8.2, time runs from left to right.

incoming gluons emits a second gluon, which is then absorbed by the other. One may verify that these are indeed distinct possibilities. In Figure 8.3(c), we show a third and more complicated possibility that builds on what is happening in Figure 8.3. Here, the incoming gluons exchange a pair of gluons, emitted and absorbed at different times. Finally, Figure 8.3 shows something else that can happen: the two incoming gluons can somehow fuse at a point, such that the two final-state gluons are immediately produced.

These diagrams are called *Feynman diagrams* and give us a very handy way to think about a very complicated process (the scattering of quantum gluons!). Looking at the diagrams in more detail, we can see a number of different features. First, each diagram must have two curly gluon lines that come in from the left, and two that go out to the right. These are called *external lines* and correspond to the fact that we have chosen to consider what happens when two gluons scatter to produce two other gluons. But there are also *internal lines*, such as the horizontal line in the middle of Figure 8.3(a) or the exchanged gluon lines in Figures 8.3(b) and 8.3(c). Whereas external lines represent particles that come in from the initial state, or emerge into the final state, internal lines represent particles that are never seen directly, and which have a curious physical interpretation. In the quantum theory, it is almost as if we can somehow "borrow" energy from the universe to make these intermediate particles, provided we give it back in some suitable time, so that they disappear again. Such particles are thus always emitted from one place, and absorbed somewhere else, which is precisely what we see in Figures 8.3(b) and 8.3(c). Another thing we can notice about these diagrams is that they contain points – called *vertices* – at which the gluons meet. Figures 8.3(a)–8.3(c) contain only vertices in which three gluons meet at a point, whereas Figure 8.3(d) has a four-gluon vertex. In fact, the equations of Yang–Mills theory tell us that these are the only possibilities.

If the only use of Feynman diagrams was to help us visualise quantum scattering processes, this would already be incredibly useful. But they are a lot more powerful than this. It turns out that there are very precise rules – called *Feynman rules* – that we can use to convert each Feynman diagram into a mathematical contribution to the scattering amplitude for our process of interest. That is, there are literally bits of algebra that we can associate with each external line, internal line, or (type of) vertex, which act as building blocks for the full scattering amplitude itself. Furthermore, the rules of QFT tell us that whenever we have more than one possible diagram connecting a given initial state with a given final state, we must take the mathematical contribution from each diagram individually and simply add them together. All of this is easier said than done, but we at least have a completely systematic way of calculating the scattering amplitude for a given process. We first draw all allowed Feynman diagrams, where the particular QFT we are working with tells us what the different vertices are that we can use. Then we convert each diagram into a mathematical contribution to the scattering amplitude. Finally, we sum all the diagrams together, to get the complete amplitude. It should be stated that Feynman diagrams are not the only way to calculate amplitudes – indeed, for many practical calculations, such methods have been superseded by faster approaches. However, the language of Feynman diagrams and rules is perhaps the simplest for understanding what we need for this book, and thus we shall continue to talk about them.

Looking again at the particular examples in Figure 8.3, we can see that not all diagrams with the same initial and final states have the same number of vertices (points at which the gluons meet). Figures 8.3(a) and 8.3(b) have two vertices, whereas Figure 8.3(c) has four. When we convert each diagram into a mathematical contribution to the scattering amplitude, those with more vertices usually contribute a smaller amount compared to those with less. Thus, to a first approximation we can take only those diagrams with the fewest number of vertices for a given number of external lines. As we add more and more vertices, this corresponds to adding ever smaller corrections to our approximate result for the scattering amplitude. It is in this sense that Feynman diagrams and rules provide a systematic way to improve our approximations to scattering amplitudes in quantum field theory. However, the mathematics of diagrams with more vertices gets increasingly complicated so that we are often forced to stop adding vertices past a certain number.

8.4 The Problem with Gravity

On the face of it, the above methods seem to be applicable to any type of field theory. We have seen that all matter and forces in nature can be described by fields and that these have wave-like solutions. What we mean by "particles" are little lumps or *quanta* of these fields, and as soon as we have such particles, we can think of forming Feynman diagrams and rules. So what goes wrong when we try to do this for gravity? The answer lies in a crucial aspect of calculating the scattering amplitude that we glossed over above. We said that one must add together all diagrams that lead to the same final state. However, even if we focus on some diagrams individually, it turns out that they correspond to different possibilities. In particular, let us look at Figure 8.3(c). In this case, two gluons are exchanged between the incoming particles. When we analyse the mathematical properties of such diagrams using the rules of QFT, it tells us that the energies and velocities of the exchanged gluons are not completely determined, such that we must sum over all possibilities. What creates this complication, when we consider the mathematics of what is going on, is that Figure 8.3(c) has a *loop* in it: the two exchanged gluons create a square in the diagram, such that starting at any one corner, we can travel round the entire square and end up at the same place. The mathematics inherent in the Feynman rules tells us very clearly that every time we have such a loop, we have to sum up over the energy that can be "flowing around the loop". To be extra pedantic, I can tell you that it is the number of *independent* loops in each diagram that matters. To illustrate this, I have given an example of a two-loop diagram shown in Figure 8.4. I can think of there being three loops in this diagram: the two individual squares and also the larger loop shown in red. However, the latter can be made by combining parts of the two smaller square loops, such that we should not consider it as independent from the others. In line with the above comments, the rules of QFT in this case tell me to sum over two sets of energies, that can flow around each loop.

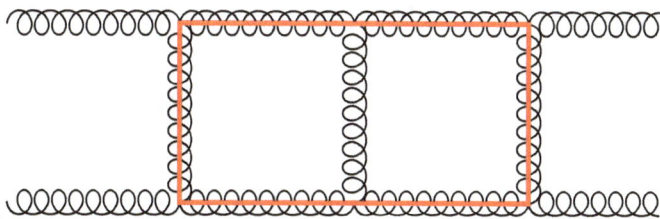

Fig. 8.4.   A two-loop diagram for gluon scattering. The loop shown in red is not independent of the two individual square loops.

As mentioned above, internal lines correspond to weird quantum particles where we have somehow borrowed a certain amount of energy from the universe, which we must return at some point to balance the books. But this can be any amount of energy, such that in summing over all possible energies that can flow around each loop in a Feynman diagram, we must include potentially infinite amounts of energy. This is in stark contrast to the external lines of Feynman diagrams that correspond to particles we can actually see and which can then have only a finite amount of energy. Perhaps unsurprisingly, if we ask what the contribution to the scattering amplitude is from diagrams containing loops, the answer is infinite! This appears to make QFT immediately useless, given that we are supposed to be using it to calculate the likelihood of a particular final state occurring. One of the first things we know about chances of things happening is that you cannot have more than a 100% chance. You certainly, therefore, cannot have an infinite chance of anything occurring.

This thorny issue disrupted the history of QFT for quite some time in the early to mid-part of the twentieth century. However, our modern way of thinking is that we should not be surprised by this behaviour. We have made a potentially faulty assumption in performing calculations involving Feynman diagrams with loops: we have included the effect of particles running around the loops that have infinite energy, which implies that we can trust our theory to work for particles with very high energies. This is almost certainly untrue, and it may well be the case that our QFTs break down at some point, to be replaced by some more underlying theory. In that case, we should be able to somehow "ignore" the very high energy particles, in a way that still allows us to predict results at lower energies, that are accessible in contemporary experiments. We can indeed do this, and the very non-obvious consequence is that we can choose to get rid of the infinite answers in Yang–Mills theory by redefining the strength of the strong force. That is, we can think of the strength of the strong nuclear force as depending on the energy of the gluons that are interacting. The absolute strength of the force is not predicted by the theory, but how it changes with energy is. Indeed, such calculations are used everyday in comparing theory with data at the Large Hadron Collider and related experiments. This idea is called *renormalisation*, which is a fancy word for changing the value of something. In this case, we have adjusted the strength of the strong force, which we can also think about in terms of changing the mathematical expressions for the three- and four-gluon vertices that enter Feynman diagrams. In the Standard Model more widely, particle masses and the strengths of the other forces will also get renormalised.

Now let us consider General Relativity. As in Yang–Mills theory, we can set up the language of Feynman diagrams and rules, and draw diagrams such as those appearing in Figures 8.3 and 8.4, but where the gluons are replaced by gravitons. Now, however, it turns out to be impossible to get rid of the infinite answers for Feynman diagrams, by redefining the strength of the gravitational force. As we include diagrams with more and more loops, we must add more and more Feynman rules to the theory, such that it never gives finite answers that can be compared with the experiment. Some progress can be made by simply imposing that there is a maximum energy that can be associated with internal lines. However, this can never be a truly underlying description of gravity, as this arbitrary energy must be covering up for the fact that there is some underlying theory that replaces General Relativity above this energy.

If you find it difficult to follow the above discussion, do not worry. The short version is that calculations in QFT give answers that we can ultimately make sense of for the Standard Model forces but not for gravity. As to why this is, there are several possibilities. First of all we know that, at least in our traditional way of thinking, gravity is simply of a different nature to the other forces. The Standard Model lives in spacetime, whereas – as we saw in Chapter 7 – gravity is *spacetime itself*. Might this somehow lead to some sort of inconsistency when we try to include quantum effects? If so, how we are to reconcile this with the fact that quantum effects are apparently needed to rescue General Relativity so that it can say something about the centres of black holes or the moment of the Big Bang? Another possibility is that GR might be OK as a theory after all. Rather, it is our way of doing quantum field theory – which involves certain approximations – that is the problem. Several people have looked at formulating the quantum theory exactly, using supercomputers to carry out the necessary calculations. However, the results are not yet conclusive.

The next possibility we can consider is that there is indeed a quantum field theory of gravity, but that it is not GR. One way of examining this is to start with General Relativity and to add extra particle content. It is known that adding certain particles – whose properties are very closely related to those of the graviton – can cancel out the effects of Feynman diagrams that would otherwise cause problems. The underlying reason for this is that such theories turn out to have extra mathematical symmetry properties, over and above those found in GR itself. One such theory has the highly exotic name of $\mathcal{N} = 8$ supergravity, and a debate has raged for around 40 years about whether it might be a well-behaved gravity theory or not. At the time of writing, it is thought that its nice mathematical properties might break down for Feynman diagrams with seven (!) loops, and progress is being made towards testing this directly.

Finally, there is the possibility that there is a well-behaved quantum theory of gravity but that it is not a quantum field theory. One such candidate theory is called loop quantum gravity, which is sadly beyond the scope of this book. We will have cause to mention another candidate theory – string theory – later on.

We have here seen that gravity is apparently very different to the other forces in nature, in that it refuses to be treated using similar methods. This casts doubt on whether we can ever reconcile the very different theoretical frameworks of quantum field theory and General Relativity and certainly leads us to believe that there cannot be any kind of deep connection between the field theories of Standard-Model-like forces and that of gravity. That there is such a connection is true, however, and we start to explore this remarkable idea in the following chapter.

Summary

In this chapter, we have seen the following ideas:

- A key quantity in quantum field theory is a *scattering amplitude*, which is related to probabilities for particles to interact.
- Everything measured by particle accelerator experiments, such as the Large Hadron Collider, can be related to scattering amplitudes.
- The different ways particles can interact when they collide can be represented using *Feynman diagrams*. Each diagram can be converted into a precise mathematical contribution to the scattering amplitude, using *Feynman rules*.

- Feynman diagrams can have loops in, such that every time there is a loop, we must add up the contributions from all possible energies of the internal particle lines.
- This leads to infinite answers for scattering amplitudes, which makes no sense. In the Standard Model, we can cure this problem by redefining the strengths of the forces, and masses of the particles.
- In gravity, the infinite answers remain, which forces us to consider that a quantum version of GR by itself does not make sense.

Chapter 9

The Double Copy: Gravity from Gluons

In the previous chapter, we saw how scattering amplitudes allow us to compare our (quantum) field theories of fundamental particles with particle accelerator experiments. They are also highly useful quantities from a conceptual point of view. Decades of experience in studying scattering amplitudes has taught us that they go right to the heart of quantum field theory. The mathematical form of scattering amplitudes in various theories contains intricate structures that have in turn taught us new ways of looking at field theories and of calculating with them. In recent years, this idea has been extended to finding relationships between *different* theories, revealing that these are much more closely related than previously thought possible. In this chapter, we get our first glimpse of this, by showing how the theory of gluons turns out to be extremely closely related to gravity, if thought about in the right way.

9.1 Feynman Rules for Gluons and Gravitons

In order to motivate the relationship we will uncover, let me first give some more details about what scattering amplitudes actually look like in the theory of gluons. To do so, I will break one of the most established rules of writing science books of this type, namely that they should contain no equations. Admittedly, I have broken this rule already in Chapter 4, in writing Einstein's equation $E = mc^2$.

But that equation is so absurdly famous that I was assuming you had seen it before, whether or not you know how to interpret it. Here, I want to show you some much more complicated equations, but I will hopefully be very clear about why I am doing so. Indeed, for the point I want to make, I genuinely believe that it will help you if you *do not know* what any of the symbols mean! In any case, here goes.

We saw in the previous chapter that there are specific mathematical expressions associated with every distinct feature of a Feynman diagram, e.g. internal lines, external lines, and vertices where the gluons meet. Let me now show you what the Feynman rule for three gluons actually looks like:

$$-gf^{abc}\left[(p_1 - p_2)_{\mu_3}\eta_{\mu_1\mu_2} + (p_2 - p_3)_{\mu_1}\eta_{\mu_2\mu_3} + (p_3 - p_1)_{\mu_2}\eta_{\mu_3\mu_1}\right].$$

Anyone expert enough to know what these symbols mean is almost certainly not reading this book, so I will assume you are entirely ignorant and allow you to feel smug if this is not the case. To either the uninitiated or experts regarding gluons, the formula looks complicated. But we can at least say that it can be typeset on a single line so that we can actually imagine someone doing some sort of calculations with it.

Now let me show you the analogous result in General Relativity, namely the Feynman rule for three gravitons meeting at a single point:

$$\text{Sym}\left[-\frac{1}{4}P_3(p \cdot p'\,\eta^{\mu\nu}\eta^{\sigma\tau}\eta^{\rho\lambda}) - \frac{1}{4}P_6(p^\sigma p^\tau\eta^{\mu\nu}\eta^{\rho\lambda})\right.$$

$$+\frac{1}{4}P_3(p \cdot p'\,\eta^{\mu\sigma}\eta^{\mu\tau}\eta^{\rho\lambda}) + \frac{1}{2}P_6(p \cdot p'\,\eta^{\mu\nu}\eta^{\sigma\rho}\eta^{\tau\lambda}) + P_3(p^\sigma p^\lambda\eta^{\mu\nu}\eta^{\tau\rho})$$

$$-\frac{1}{2}P_3(p^\tau p'^\mu\eta^{\nu\sigma}\eta^{\rho\lambda}) + \frac{1}{2}P_3(p^\rho p'^\lambda\eta^{\mu\sigma}\eta^{\nu\tau}) + \frac{1}{2}P_6(p^\rho p^\lambda\eta^{\mu\sigma}\eta^{\nu\tau})$$

$$\left.+P_6(p^\sigma p'^\lambda\eta^{\tau\mu}\eta^{\nu\rho}) + P_3(p^\sigma p'^\mu\eta^{\tau\rho}\eta^{\lambda\nu}) - P_3(p \cdot p'\,\eta^{\nu\sigma}\eta^{\tau\rho}\eta^{\lambda\mu})\right].$$

We can immediately see that this is a lot more complicated than the gluon case. There are even more types of weird symbols than the thing we had above, which already looked bizarre. Let me also tell you that I have quoted this result from a classic 1967 paper by

Bryce de Witt,[1] one of a series of three papers in which the quantum version of General Relativity was spelled out in painstaking detail. The author used a very clever notation in writing the above result that makes it look simpler than it actually is! If an expert were to expand the notation we have written, then the rule for three gravitons would have 171 distinct bits or "mathematical terms" in it. For comparison, the gluon result has 6. I need not explain to you that 171 is a lot bigger than 6, and thus the theory of quantum General Relativity is stupendously more unwieldy than the quantum theory of gluons.

In fact, the problem gets worse. As we saw in Chapter 8, we can also have four gluons meeting at a point. The Feynman rule for this again turns out to have six terms in, albeit of a different nature to the three-gluon result. In gravity, the four-gluon vertex rule has 2580 (!) terms, which the late great Bryce de Witt managed to reduce to 28 in his original paper, by grouping things together in a particularly ingenious mathematical way. In the theory of gluons, there are no vertices involving higher numbers of gluons, and thus there is some sort of limit to how complicated the calculations can get. In gravity, however, there are vertices involving higher and higher numbers of gravitons, all of which have to be included as we include more and more effects in Feynman diagrams. Thus, the huge complication we have already seen is merely the tip of a colossal iceberg.

There are two messages that I want you to take in from this discussion. First, the traditional way of thinking about quantum calculations in gravity, in terms of Feynman rules and diagrams, is so ridiculously cumbersome that at some point the calculations literally become impossible. In fact, as a member of the generation of scientists who grew up being able to ask computers for help with their calculations, I find it astonishing that de Witt and his contemporaries in the 1960s were able to say anything at all about quantum gravity! Or, even if they did, that they got their calculations right and able to agree with each other. Computer power has itself extended our ability to chip away at calculations in gravity, but they remain enormously challenging, and this itself held back research into possible theories of quantum gravity for decades.

---

[1]For those that like proper-looking academic references, this can be found in *Phys. Rev.* 162 (1967), 1239–1256.

The second point I want to make is that the methods discussed earlier seem to tell us that gravity is completely different to the theory of gluons. That is, even if we are able to assemble all the mathematics together to calculate scattering amplitudes in General Relativity, the fact that things are so much more complicated in gravity means that the results will look nothing at all like the corresponding amplitudes in the theory of gluons. To make this idea of comparing scattering amplitudes more formal, we can use a conventional way to characterise them. First, we can talk about the number of external particles that they have (i.e. the number of incoming particles, plus the number of outgoing particles). This does not completely fix the type of Feynman diagram, however, as we can see by comparing Figures 8.3(a) and 8.3(c). Both of these diagrams have four external particles, but Figure 8.3(c) has more going on. A succinct way of saying this is that it has a loop, and this distinction also applies to Figure 8.4, which has two loops. Another way of talking about Feynman diagrams is in terms of how many vertices they have, and by studying Figures 8.3 and 8.4, we can clearly see that the number of vertices increases as the number of loops increases. In fact, one may show mathematically that if we specify both the number of external particles and the number of loops, this completely fixes the number of vertices in a given diagram. Hence, it is common to talk separately about the scattering amplitude for a process where we have isolated a given number of external particles and loops. By doing this both in gluon theory and gravity, we can then put the results side by side and see how they compare. It is this comparison that, until relatively recently, showed us nothing particularly useful. All of this changed dramatically in 2010, as we now explore.

9.2   The Double Copy

In 2010, a paper appeared by a team of researchers (Zvi Bern, John Joseph Carrasco, and Henrik Johansson) split between Los Angeles and Paris. The title was *Perturbative Quantum Gravity as a Double Copy of Gauge Theory*, and the paper suggested a remarkable relationship between scattering amplitudes in theories such as those describing gluons ("gauge theories") and possible field theories of quantum gravity. As so often in science, this development did not happen in isolation, but built upon previous work by the same authors, even earlier work by the first author, and yet more earlier work in the mysterious "string theory" that was briefly mentioned at the end of the previous chapter. We describe the latter in the following, but to do so right away would take us away from the field theories we are talking about at present. To understand the idea, I am going to have to break once more the rule of not showing you any equations, but my own special rules will still apply. In particular, I am going to continue to believe that not knowing what the symbols mean actually helps you, because I want you to play the traditionally fun game of *spot the difference*. The two objects I want you to compare are the mathematical formulae for scattering amplitudes in the theory of gluons (Yang–Mills theory) and gravity (meaning General Relativity, plus possibly some extra things to be clarified in due course). As described in the previous section, I will take the scattering amplitude that arises from Feynman diagrams with a given number of external particles and a given number of loops. First up, here is the result for gluons:

$$g^{m-2+2L} \sum_i \int \prod_{l=1}^{L} \frac{d^D k_l}{(2\pi)^D} \frac{1}{S_i} \frac{n_i c_i}{\prod_{\alpha_i} p_{\alpha_i}^2}. \qquad (9.1)$$

Maths is, of course, terrifying, but please don't be frightened. I am showing this equation as a piece of art rather than of science, given that I am not asking you to understand its content directly. Without further ado, here is the result for the similar scattering amplitude in gravity:

$$\left(\frac{\kappa}{2}\right)^{m-2+2L} \sum_i \int \prod_{l=1}^{L} \frac{d^D k_l}{(2\pi)^D} \frac{1}{S_i} \frac{n_i n_i}{\prod_{\alpha_i} p_{\alpha_i}^2}. \qquad (9.2)$$

Having given you both results, we can now look carefully at the two formulae – which we can view simply as pictures – and try to find the differences between them. You may wish to give yourself some time to do this yourself, before reading my answers in the following paragraph.

I think there are only two differences between the two formulae, although it depends somewhat on how you count them:

(i) The symbol g in Equation (9.1) has been replaced by the symbols

$$\left(\frac{\kappa}{2}\right)$$

in Equation (9.2). Of course, you may have counted this as more than one difference, depending on how you have counted the separate symbols κ, 2 and the round brackets. That is perfectly allowed, as I should have been more precise in specifying what I meant by a "difference".

(ii) The symbol c_i in Equation (9.1) has been replaced by the symbol n_i in Equation (9.2). Interestingly, this same symbol n_i *already* appears in Equation (9.1) so that it is common to both the gluon and gravity theories.

At this point, we know nothing about how to interpret these formulae. And yet, if you believe my claim that they represent, respectively, comparable scattering amplitudes in gauge theory and gravity, then we can see the following idea: *if we write scattering amplitudes in the right way in the theory of gluons, we can perform an apparently minor tweak and immediately get a scattering amplitude in gravity.* Given our traditional methods described earlier, this is an astonishing miracle that runs completely counter to our expectations. We saw, did we not, that calculations in gravity are enormously more complex than those in the theory of gluons. However, we saw this complexity at the level of individual Feynman rules. What happens when we translate entire Feynman diagrams into algebra, and then add them all together, is that an amazing simplification occurs. The results in gravity, if thought about in the right way, really do look like they can be obtained directly from a much simpler theory of gluons, rather than gravitons!

Let us now look at the formulae in more detail, to which end I will try to describe as best I can what various parts of them represent. I will not give you a full explanation – for my university students, that takes around five years. But we can at least get a feeling about where different bits of the equations come from and thus what the "tweak" that takes us from Equations (9.1) to (9.2) actually corresponds to. Perhaps the easiest thing to explain is difference (i) above, namely the replacement of the symbol in front of the entire formula. In any given theory of a force field, there is a number that represents the strength of the force. In the theory of gluons, we typically call this number g for some reason, whereas for gravity, we call it κ divided by two, where the "divided by two" is there for weird historical reasons. The number κ can in turn be related to the strength of the gravitational force as it arose in Newton's original theory of gravity. In simple terms, then, replacement (i) above means the following: *replace the strength of the strong force, with the strength of gravity*. This makes perfect sense if we are trying to make a gravity scattering amplitude out of a gluon one!

More subtle is the second replacement, for which we are going to have to say some words about what the other symbols mean. First, let us say that if gluons come together and interact, each gluon will have various properties that characterise its motion, such as a velocity, energy, and so on. We call such quantities *kinematic*, to distinguish them from the other types of properties that particles have. For example, electrons carry electric charge, and quarks carry colour charge in addition, where the latter is the type of charge that the strong nuclear force "sees". Gluons are the particles that transmit the strong force, but they can also carry colour charge themselves. If they didn't, then they wouldn't be able to interact with each other, as they can only "see" particles that carry a colour charge. In general, the scattering amplitude will depend both on the kinematic information (how the gluons are moving through spacetime) and the colour charge degrees of freedom. In Equation (9.1), the symbol c_i denotes the part of the formula that depends only on the colour information, whereas the n_i symbol depends only on the kinematic stuff. We needn't talk too much about the rest of Equation (9.1), but in case you are interested: the funny symbol Σ out the front tells us to sum over all possible diagrams, and the strange $\int$ symbol with various things to the right of it tells us to sum up over the various energies running around the loops, as we said was needed at the end of the previous chapter. Finally, the things at the bottom of the formula turn out to arise from the Feynman rules for the internal lines of each diagram and end up being the same in both the theory of gluons and gravity.

Having now told you, at least vaguely, what the n_i and c_i symbols represent, we can look again at Equation (9.2) and see that replacement (ii) above amounts to stripping off the c_i from Equation (9.1) and replacing it with a second symbol n_i, such that two of these

appear in Equation (9.2). In words rather than symbols, we must
get rid of the part of the scattering amplitude in gluon theory that
depends on the colour charges and replace it with a second set of
kinematic information. This makes some sort of sense, in that there
are no colour charges in gravity. Instead, gravitons "see" mass and
energy. These are kinematic things, i.e. they are to do with motion.
Thus, we expect to have to get rid of colour information and to have
only kinematic information in gravity. The big surprise to physicists
in 2010 was that this happens quite so simply and elegantly, for
Feynman diagrams with *any* number of external particles and loops.

Quantum field theory has been around, in some form, since the
early twentieth century, as has General Relativity. So it is natural to
ask why this odd, yet extremely precise, relationship between gluon
and gravity theories was only noticed as recently as 2010. One aspect
that is important is that it is not generically true that formulae such
as Equation (9.1) lead to simple gravity formulae like Equation (9.2).
It turns out that the precise mathematical form of the n_i symbols in
Equation (9.1) is not unique, but can be chosen in different ways, so
that the total scattering amplitude remains the same. It is only for
very particular choices of these symbols that one may make the above
replacements to get a gravity amplitude, and this choice corresponds
to a very special relationship between the symbols c_i and n_i in the
gluon theory. This mysterious property was discovered slightly earlier
in 2008 and is known as *BCJ duality*, after its inventors Bern, Carrasco, and Johansson, who we already saw earlier. In lay terms, what
they found is that the kinematic information in scattering amplitudes
must be intricately related to the colour information, in a very precise way. To give an analogy, we may imagine our gluons as actors
moving around on a stage, where the latter represents spacetime.
The different colour charges that the gluons can have can then be
thought of as being differently coloured costumes that the actors are
wearing. Why, however, should the colour of their jumper influence
how they move around on the stage?[2] BCJ duality tells us that there

---

[2]In Peter Greenaway's film *The Cook, the Thief, his Wife, and her Lover* (1989),
the actors' clothes change colour based on which room they are in. This is perhaps
the closest that art has gotten to representing the physics of BCJ duality, pre-
sciently doing so almost twenty years before BCJ duality appeared in the physics
literature!

must nevertheless be some sort of relation of this type. Despite some progress, this mysterious linking between colour charges and kinematic information is still not fully understood at the time of writing. In my view, it is likely that some new mathematical way of thinking about the theory of gluons is needed, in order to make this previously hidden structure show up explicitly.

Provided we have chosen our symbols n_i carefully enough, Equation (9.2) does indeed follow from Equation (9.1) and is a correct gravity scattering amplitude, as can be verified by other means. What has been found, then, is an extremely efficient procedure for recycling results from the theory of gluons to make gravity results that completely bypasses our traditional methods of calculating in gravity! There is a special name for the earlier procedure that already appears in the title of the discovery paper mentioned earlier. It is called the *double copy*, and the origin of this name is that starting with Equation (9.1), one must take two copies of the symbols n_i in forming Equation (9.2). Used properly in a full sentence, we would say the following: *the double copy turns gluon scattering amplitudes into graviton ones.*

So far we have talked about the double copy as if it is actually true, so before carrying on, we should address this. The double copy – and the corresponding link between colour and kinematic behaviour of gluons of BCJ duality – is what is known as a conjecture. That is, we lack a formal proof that the double copy will work for arbitrary numbers of loops, although it is more or less established that the double copy will work provided BCJ duality is true. That is, a proof of the double copy may rely on being able to show that one can *always* choose the symbols n_i in the "right" way, regardless of how many loops there are in our Feynman diagrams. For Feynman diagrams with no loops, however, we can indeed prove that the double copy works for any number of external particles. In that case, it also corresponds to something that was already known about, as we now explore.

9.3 The Double Copy from String Theory

In Table 6.1, we motivated quantum field theory as being the theory that includes the effects of both Special Relativity (needed when objects move very fast) and Quantum Mechanics (needed when objects get very small). However, I refrained from pointing out that it is not the only theory that manages this. Another such theory is *string theory*, whose basic idea is simple enough to state the following: in quantum field theories such as the Standard Model of Particle Physics, the fundamental objects in nature are *particles*: tiny point-like things that have no additional structure in themselves. One may instead consider a different possibility, namely that the fundamental objects in nature do indeed have a structure, and the simplest possibility is that they are little strings. Let us not worry for the moment about what these strings are "made of", in the same way that we didn't really ask what the fields in the Standard Model were made of. They are simply *there*, such that we regard the stuff around us as made of fields, rather than the fields being made of something else.

If we consider strings, a moment's thought will establish that there are at least two different types of strings that one can consider. These are shown in Figures 9.1(a) and 9.1(b) and are known as *open* and *closed* strings, respectively. If you fancy playing around with a ball of wool, you can clearly make such configurations yourself. The open string is what we would call a standard piece of string, with two end points. You can then make a closed string by gluing the ends together. This analogy is slightly flawed only in that, in the basic assumptions of string theory, there is no need for a separate "glue substance" to make a closed string. These strings simply exist. Furthermore, there is no need to consider more complicated configurations of string, such as knots, intertwined loops, or entire knitting patterns. These knots and loops will always pass through each other, leaving only the possibility that a given string is either open or closed.

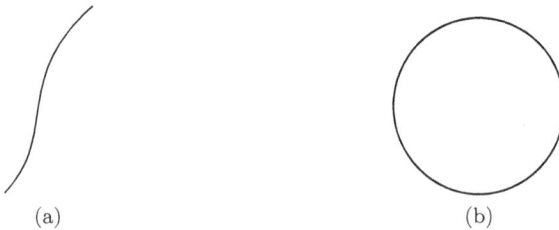

(a)   (b)

Fig. 9.1.   (a) An open string; (b) a closed string.

Since the 1960s, it has been known how to write down a theory of strings. These are very small and moving very fast in principle. Thus, this theory includes the effects of both Quantum Mechanics and Special Relativity. The original motivation for this theory was to try and understand the strong nuclear force. During the 1960s, a large number of strongly interacting particles were discovered that were given the name of *mesons*. The way these particles moved (in particular how they were rotating) seemed to indicate that they might correspond to quantum, relativistic open strings, where the strength of the strong force holding these objects together would correspond to the "tension" in the string, in the same way that we can think of ropes in everyday life as having a certain tautness. This hypothesis was disproved by experiments in the 1970s, which firmly established that there were point-like particles living inside mesons, which we now understand to be quarks and antiquarks. The modern understanding of the string-like behaviour is then that this is what the gluon field that binds the (anti)quarks together happens to be doing. However, this field really corresponds to particles (gluons) rather than strings, and thus the original string explanation of the strong force cannot be true.

Despite this initial setback, string theory continued to be studied and was spectacularly reinvented in the 1980s as a potential theory of quantum gravity. To motivate this idea, let us first note that all of our past and current collider experiments have taught us that nature looks very much like a quantum field theory. Such theories contain particles rather than strings, and thus it must be true that string theory, if it exists in nature, should somehow kick in at very high energies, reducing to a quantum field theory at low energies. This is similar to how, in Table 6.1, QFT reduces to other theories under certain conditions. One way to think about lowering the energy is that we are somehow zooming out from the strings, such that we can't notice the highly energetic processes that are going on inside them. Feel free to literally do this by placing this book, with Figure 9.1 displayed, on the other side of the room from where you are sitting. As the book gets further away, you will at some point not be able to resolve the details of the figure, and the strings – either open or closed – will look like particles!

If strings at low energy look like particles, we can ask what type of particle they correspond to. In the original string theory, the answer is very intriguing indeed: open strings give rise to gluons at low energy, and closed strings give rise to gravitons! More precisely, one may show that string theory reduces to a quantum field theory at low energy, where one obtains Yang–Mills theory from open strings and a variant of General Relativity from closed strings. The only complication in the latter is that there are two additional particles that accompany the graviton. They are sometimes called the *axion* and *dilaton*, and the complete theory in which they join the graviton is known as $\mathcal{N} = 0$ supergravity. For our purposes, "GR plus extra bits" will suffice.

String theory caused immense excitement from the 1980s onwards, as it finally seemed as if we had a theory in which all of the fundamental forces of nature fit into the same overarching theoretical framework. Not only this, but there was a sense of inevitability, in that gravity was not optional but had to be there. However, our quick summary of string theory has glossed over its many baffling features. The original theory only makes mathematical sense in 26 spacetime dimensions, in obvious contradiction with the fact that we apparently see only 4 (3 space dimensions and 1 of time). Critics regard this as an obvious failure of the theory, whereas enthusiasts point out that the theory instead *predicts* how many spacetime dimensions our universe must have and devise elaborate – but oddly plausible – schemes for how these dimensions might be curled up so that we cannot see them. The original theory does not have a way to generate particles like the matter particles we observe (e.g. quarks and electrons). When this is amended – by generalising the mathematical description of what we mean by a string – then the theory requires 10 spacetime dimensions and reduces to different field theories at low energy. Open strings give rise to $\mathcal{N} = 4$ Super-Yang–Mills theory, which can be thought of as a theory of gluons, interacting with extra matter particles whose properties are highly constrained. Closed strings give rise to $\mathcal{N} = 8$ supergravity, which is the theory that we briefly mentioned at the end of the previous chapter as a potentially well-behaved field theory of quantum gravity. Neither of these theories is obviously manifest in the universe we live in, although this is not necessarily a problem given that we know how these theories might reduce to the theories that we do have at low energy.

Further excitement was generated in the 1990s when it was realised that string theory is not only a theory of strings. If we think of strings as being "one-dimensional", meaning that we need a single number to tell us how far along a string we are, there are also higher-dimensional objects called D-branes. String theory thus includes an exotic zoo of different types of objects, all of which may be present in our universe at its smallest scales. However, due to this rich diversity, plus the fact that there are so many ways of curling up the extra spacetime dimensions, it is simply not known – or apparently easily knowable – how the Standard Model of Particle Physics may arise from such an underlying description. For this and other reasons, string theory has gotten an astonishing amount of bad press over the last few years, culminating in a series of very public spats in the 2000s known as the "string wars". It may seem odd that such levels of anger can be generated by esoteric questions regarding the nature of reality, but such bickering has always been a part of science, and one must remember that scientist's livelihoods depend on which particular unproven fundamental theory may be flavour of the month.

I consider myself unbiased when it comes to the merits or, otherwise, of string theory as a fundamental theory of nature. What I do see very clearly, however, is that it acts as an incredibly useful mathematical edifice that ties together the various theories that we do know are there in nature. One such use is an elegant confirmation of the double copy between gluon theory and gravity. Let us take the original string theory alluded to earlier, whose low energy limit gives us gluons (open strings) or gravitons plus extra stuff (closed strings). Now let us note that, in string theory, we can talk about the scattering of strings much as we talk about the scattering of particles in a quantum field theory. Unlike Figure 8.2, it is not quite true that we can represent incoming or outgoing strings by lines. Consider taking a particle at some point in space and watching it move with time. At each given point in time, there will be a definite point at which the particle is located so that it indeed traces out a line as time evolves, as shown in Figure 9.2(a). In physics parlance, this is called a *worldline,* and this concept first gained popularity when Special Relativity came along. If we now consider strings instead, then the entire string must be plotted at each given point in time. As time passes, such objects will trace out a so-called *worldsheet,* and examples for open and closed strings are shown in Figure 9.2(b). Note that an open string worldsheet looks like a ribbon, whereas a closed string worldsheet looks like a tube. They are thus different types of

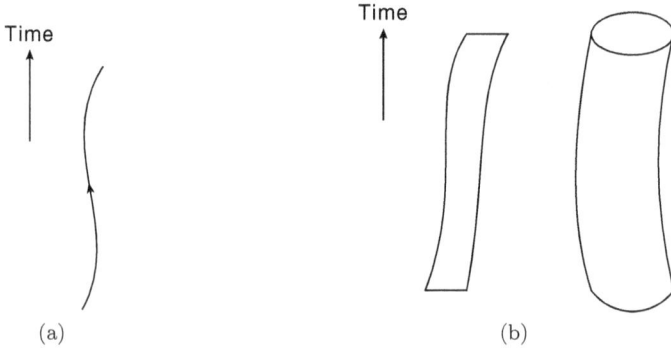

Fig. 9.2. (a) As time passes, a particle traces out a *worldline* in spacetime; (b) strings instead trace out *worldsheets*.

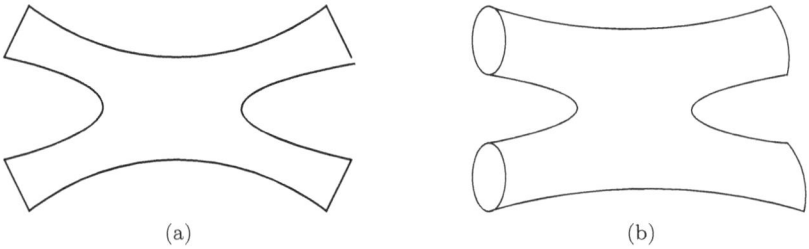

Fig. 9.3. Example Feynman diagrams for the scattering of (a) open strings; (b) closed strings.

mathematical objects, and these worldsheets should take the place of the lines in Figure 8.2 when we talk about string, rather than particle, scattering events. Once we do this, we can happily draw Feynman diagrams for the scattering of strings. Examples for open and closed strings are shown in Figure 9.3. Note that neither of these diagrams has a loop in it. If you are curious as to what this looks like for string diagrams, an example of a loop diagram is shown in Figure 9.4.

Returning to stringy Feynman diagrams that have no loops in, it was shown in a seminal paper in 1986 that one can write a very precise mathematical relation between scattering amplitudes for closed strings and comparable amplitudes for open strings. The authors were Kawai, Lewellen and Tye, and thus these relations became known as the *KLT relations*. Given that they must be true for strings of any energy, one may see what these relations imply at low energies, where the string theory becomes a quantum field theory. Then, given

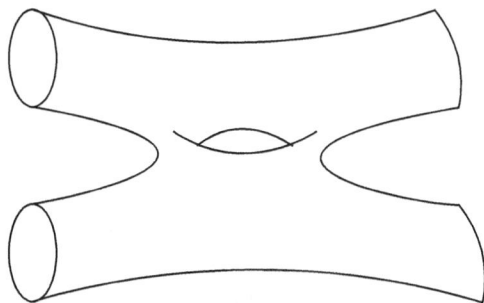

Fig. 9.4.   Example Feynman diagram with a loop in string theory.

that open and closed strings give us gluons and gravitons respectively, we must end up with a relation between gluon and graviton scattering amplitudes. This turns out to be the double copy, and string theory thus provides a derivation of the double copy, if we consider only those Feynman diagrams that have no loops.

What is nice about this argument is that it is completely irrelevant whether or not string theory actually exists in nature. Here it acts as a mathematical bridge that tells us to look for relations between two field theories (gluons and gravitons) that we know are there in nature. For me personally, such arguments more than justify the development and continued interest in string theory, such that it has very much become part of the varied toolkit of the modern theoretical physicist. Unfortunately, it is not known whether the stringy derivation of the double copy works when loops are involved. Thus, I personally do not feel it is justified to say that the KLT relations are the same as the double copy, although this seems very likely. I note further that progress on generalising the KLT relations to include Feynman diagrams with loops in them has been made by Stefan Stieberger at the Max Planck Institute for Physics in Munich.

9.4   There Are Many Double Copies

In the previous sections, we have talked specifically about the double copy relating pure Yang–Mills theory (the theory of gluons interacting by themselves) and $\mathcal{N} = 0$ supergravity, which is General Relativity plus a couple of extra particles. We also saw that the approximate string theory explanation for the double copy related two different theories. There was a theory built on the theory of gluons ($\mathcal{N} = 4$ Super-Yang–Mills theory), and a theory built on General Relativity, but with a lot more particles in it ($\mathcal{N} = 8$ supergravity). This begs the question as to whether there are even more theories that can be related by the double copy. More specifically, are there other generalisations of Yang–Mills theory and, if so, are they related to generalised gravity theories?

The answers to these questions are all yes, and we thus need a convenient way to talk about the *types* of theories that are related by the double copy. Theories built on Yang–Mills theory are typically referred to as *gauge theories*, due to the abstract mathematical *gauge symmetry* that they possess and that we reviewed in Chapter 6. There may or may not be extra particles accompanying the gluons, and so the simplest way to think about what a gauge theory is, is that it contains particles with some sort of colour charge. If I want to talk about a theory of gravitons plus other particles, I will simply call this a *gravity theory*. Then we can talk about loads of different theories at once if we say the following: *the double copy relates scattering amplitudes in gauge theory and gravity.*

What are the generalised theories we are talking about? Perhaps the first thing to point out is that it is not understood how to take a completely arbitrary gauge theory – with any choice of matter particles – and to apply the double copy to get a related gravity theory. This may or may not be possible, and what has happened in the last ten years or so is that an increasing catalogue of gauge and gravity theories has been found, all of which obey the correspondence. A particularly general family of theories is those in which extra particles are added, whose masses and interaction strengths are very closely related to the gluons themselves. This in turn means that the structure of such theories stays the same if we swap the fields around in the equations, which property is known as *supersymmetry*. We can have different amounts of supersymmetry in a gauge theory, depending on how much extra stuff we add. Thus, we typically label supersymmetric gauge theories by a number $\mathcal{N}$, which denotes the amount of supersymmetry (more specifically, the number of distinct ways in which we can swap the fields around while preserving the equations of the theory). This explains the weird notation we have seen earlier when discussing the field theories that arise from string theory, and it turns out that $\mathcal{N} = 4$ Super-Yang–Mills theory is the most supersymmetric version of Yang–Mills theory that we can make in four spacetime dimensions. It has been intensively studied for decades for all sorts of reasons, with the chief one being that the large amount of specialness in the theory means that calculations (e.g. of scattering amplitudes) are vastly simpler than they are in more physically relevant theories such as pure Yang–Mills theory. Indeed, many of the world-leading calculations used at the Large Hadron Collider – such as our most precise estimate for how to make a Higgs boson – use techniques that were first developed in the study of $\mathcal{N} = 4$ Super-Yang–Mills theory.

As in gauge theory, we can also add supersymmetry to General Relativity, meaning that we add extra particle content whose properties are very closely related to the graviton. It is again conventional to label the amount of supersymmetry by $\mathcal{N}$, where the most amount of this we can add in four spacetime dimensions leads to $\mathcal{N} = 8$ supergravity. We saw above that $\mathcal{N} = 4$ Super-Yang–Mills theory is related to this by the double copy, which suggests a possible pattern. If we have theories with less than maximal supersymmetry, e.g. $\mathcal{N} = 2$ Super-Yang–Mills theory, is it true that this double copies to $\mathcal{N} = 4$ supergravity? The answer is indeed yes, such that the amount of particle content we see in the gravity theory is directly related to what we start with in the gauge theory.

We can go even further than this. It turns out that the double copy formula of Equation (9.2) – which contains two copies of the symbols n_i from Equation (9.1) – is not the most general thing we can do. We can instead use symbols n_i and $\tilde{n}_i$ from *different* gauge theories in the top of Equation (9.2), and *still* get a gravity amplitude. Once again, the gravity theory that we end up with depends on which gauge theories we start with. For example $\mathcal{N} = 4$ Super-Yang–Mills theory combined with $\mathcal{N} = 3$ Super-Yang–Mills theory would give us $\mathcal{N} = 7$ supergravity, where we simply add the numbers.

This discussion is admittedly rather dry, but the overall point I am trying to make is that we are gradually understanding that amplitudes in a huge variety of different gravity theories can be obtained from double-copying gauge theory amplitudes. We are also able to ascertain entire families of theories (such as the supersymmetric ones), all of which obey the correspondence almost automatically.

9.5   The Square Root of Einstein

We have seen a lot of physics up to now but have finally arrived at
the moment where we can understand the title of this book! Let me
first remind you what a square root is. If we have a number – for
Example 9 – we can find another number such that, if we multiply
it by itself, we get our number of choice. You will hopefully know
that

$$3 \times 3 = 9$$

so that three multiplied by itself gives nine. We call the act of multiplying something by itself *squaring it*, and we call the number 3
the *square root* of 9. All numbers have square roots, but they are not
necessarily whole numbers. For example, the square root of 36 is 6,
but the square root of 2 is $1.41421\ldots$, and in fact goes on forever.

We saw that the gravity amplitude formula of Equation (9.2)
involves taking something (the symbol n_i) from the gauge theory
amplitude of Equation (9.1) and multiplying it by itself. It is not
true that the whole of Equation (9.2) is the square of Equation (9.1)
(n.b. I do not expect you to be able to work this out, so please take
it on trust). Rather, it is merely a bit of the formula that involves
squaring something. Nevertheless, this has led to a colloquial way of
referring to the double copy by physicists working in this area. We say
things like "gravity is the square of gauge theory", meaning that we
need two gauge theories to "make" a scattering amplitude in gravity.
Conversely, we say that "gauge theory is the square root of gravity".
If we want to be more specific about which gauge and gravity theory
we are talking about, we will say their specific names. In the case
of gravity, both "$\mathcal{N} = 0$ supergravity" and "General Relativity"
take a long time to say, especially when that time is precious during
the coffee break of a conference. Thus, it is much more common
for scattering amplitude enthusiasts to refer to General Relativity
as "Einstein gravity", or simply "Einstein" for short. This theory
(or a closely related variant) is obtained from the double copy of
Yang–Mills theory. Hence:

The square root of Einstein is Yang–Mills!

9.6 Why Is the Double Copy Important?

I have tried your patience enough perhaps by telling you what a scattering amplitude is and stating that gravity amplitudes can be obtained from gauge theory ones. It is only right that I now explain why I think this is so important. The first reason is a purely practical one. When I showed you quite how complicated calculations are in quantum gravity, I stressed that this fact itself had held back research into this subject for decades. In particular, the notion of whether certain field theories of quantum gravity (such as $\mathcal{N} = 8$ supergravity) could be well behaved had stalled, due to a lack of ability to perform the relevant calculations. The double copy circumvents this, by giving us an entirely new way to calculate scattering amplitudes in gravity. These quantities are much easier to calculate in gauge theories, and what the double copy tells us is that we can simply take our gauge theory results, provided they are written in the right way, and almost immediately obtain the results we need in gravity. This new way of calculating things has revolutionised the generation of new calculations in $\mathcal{N} = 8$ supergravity, showing that previously conjectured ill behaviour does not in fact arise. New insights may be needed to push the calculations further, but the double copy can surely be expected to be part of this.

Of course, you may question how "practical" the applications are of an idea which helps us to calculate things in theories that are so esoteric that they do not immediately apply to our everyday lives. However, the double copy works for *any* theory of gravity and not just the ones that happen to have been debated as possible well-behaved quantum gravity theories. There are many results that are needed to better understand plain old General Relativity which, as we have seen, is certainly relevant to the astrophysics of our own universe, including satellite communications. The double copy provides new ways of calculating such things, and we see more of these applications in Chapter 11.

Another major reason why the double copy is important is that it indicates an entirely new relationship between different types of physical theories. We have seen that in our traditional ways of thinking, the forces underlying particle physics (as described in the Standard Model) are very much separate from gravity. The double copy spectacularly refutes this notion and offers a tantalising glimpse of an underlying common structure sitting deep beneath all our theories. Gauge theories and gravity describe absolutely everything we have ever observed in our universe, from the scenes of our everyday lives, to the tiny particles living inside the atoms of which we are made, and to the massive clusters of galaxies and other material that make up our entire universe. Physics has long sought for a supposed *theory of everything*, in which all of our current physical theories are part of a unified framework. Indeed, string theory is one such candidate theory, and it may well be the case that this theory fully explains the double copy and where it comes from. But seeking this or any alternative explanation will necessarily involve uncovering new structures and ideas that have the potential to revolutionise our understanding of fundamental physics. Certainly there is the very strong indication that our traditional ways of thinking about, and even using, theories of fundamental physics are simply not the right ones. Better methods would make structures like the double copy much more explicit, such that all theories start to look the same. In the coming years, it is highly likely that the next generation of scientists – which may include yourself – will rewrite the textbook methods of quantum field theory that have lasted for decades.

Lest I be accused of hyperbole, let me provide what I see as more evidence that the double copy is telling us something deep about what field theories are and how they are mutually related.

9.7   The Single and Zeroth Copies

We have so far seen that the double copy relates scattering amplitudes in gauge theory, with those in gravity. But this is not the only relationship we can consider. First, we could think of running the double copy backwards so that we generate results in gauge theory from those in a gravity theory. This involves replacing the strength of the gravitational force with the relevant one for a gauge theory and also replacing kinematic by colour information. The opposite of the double copy is called the *single copy*, where we can remember that the word "double" referred to the fact that we had two lots of kinematic information (as represented by the symbols n_i) in Equation (9.2). There is only one such symbol in Equation (9.1), hence the word "single". Nice as this terminology is, it is of course not giving us anything that we didn't already know about. But the act of taking the single copy suggests that we might go further and take away the kinematic symbols n_i from Equation (9.1) completely. In doing so, as in the single copy, we should replace them with a second set of colour information. In other words, we would end up with a theory involving some sort of particle that has *two different types* of colour charges. This theory has the wonderfully peculiar name of *biadjoint scalar (field) theory*, where you can trust me that each of the words has a very precise meaning that, on a good day, manages to remind a theoretical physicist what the theory represents.

We believe that biadjoint scalar field theory cannot be a physical theory of nature by itself, due to certain undesirable mathematical properties of the equations that describe it. However, it is interesting that this theory was essentially discovered by the study of the double copy – and related work – that looked at how quantities in one type of theory can be related to those in a completely different theory. Biadjoint theory has continued to crop up in various contexts in contemporary physics research and is in some ways much simpler than the gauge and gravity theories that it is related to. This is itself intriguing, as it tells us that at least some of what happens in gauge theories and gravity – including the ones relevant for our world – is essentially inherited from a much simpler underlying theory that we would never have thought of looking for otherwise. As a result, in the past few years, I myself have become thoroughly obsessed with biadjoint scalar theory. I have literally lain awake in the middle of the night, pondering the equations of the theory over and over again in different ways, trying to work out what on earth its uninvited presence in our research papers is trying to tell us. There is also the issue of whether this theory is an approximation to a more complete theory that may indeed have a real-world application. It may well be true, for example, that some modified version of biadjoint scalar theory describes the behaviour of certain materials, although my own thoughts on this are much too speculative to deserve a mention here.

The procedure that takes us from gauge theory to biadjoint scalar theory is referred to as the *zeroth copy* in the physics literature, owing to the fact that there are no symbols n_i in the equivalent formula of Equation (9.1). We can draw a diagram summarising everything we have seen so far, which I show in Figure 9.5. All of the relationships

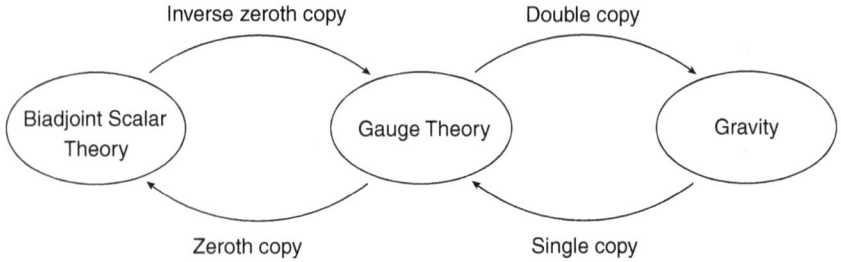

Fig. 9.5. Different types of quantum field theories are related by a family of relationships, which include the double copy. Each of these procedures allow us to take scattering amplitudes in one theory and relate them to amplitudes in a completely different theory.

we have described occur as arrows linking together different theories. To complete the diagram, we need a name for the opposite of the zeroth copy that takes us from gauge theory to biadjoint scalar theory. Different conventions exist for this, but I will follow the nomenclature of my former PhD student Nadia Bahjat-Abbas, who referred to this as the *inverse zeroth copy* in her work. Here, the word "inverse" is a mathematically precise way of saying "the opposite thing to".

My own training as a theoretical physicist consisted of spending a lot of time working on gauge theories in the first ten or so years of my research career, while at the same time being more than slightly scared of gravity. Diagrams such as Figure 9.5, therefore, make me very excited indeed. They tell me that I can apply similar methods to the theory I know and love, to things that I was previously frightened of. The diagram also tells me that there are entirely different and weird types of field theories that I would never have otherwise considered but that are somehow deeply related to things I know about. What's more, we now know that the ladder of theories in Figure 9.5 is only a subset of a much broader picture that we return to in Chapter 12. That is, there is by now a highly complex web of different types of quantum field theories, all of whose scattering amplitudes are related by correspondences, such as the double copy. Again, there is the strong suggestion that there is some common underlying structure sitting underneath all of this that, if fully understood, will give us a completely new way to think about what a field theory is. Hence my lying awake for hours on end in the middle of the night!

Summary

In this chapter, we have gotten to the key idea of this book: that quantities in entirely different theories of physics are in fact very closely related. The main points are as follows:

- Using traditional methods, scattering amplitudes (which describe how particles interact with each other) are much more complicated to calculate in gravity than in gluon theory.
- This held back progress in understanding quantum gravity for many years.
- In 2010, a new correspondence called the *double copy* was discovered. It tells us that scattering amplitudes in gravity can be straightforwardly obtained by slightly modifying corresponding formulae in gluon theory.
- More generally, amplitudes in *gauge theory* (theories related to gluon theory) are related to various generalisations of gravity (General Relativity, plus extra particles).
- The double copy follows, at least in part, from earlier work in string theory, a hypothetical theory in which particles are replaced by small string-like objects.
- But string theory is not needed to be a real theory of nature for this to work.
- Similar correspondences link gauge theories to a theory known as *biadjoint scalar theory*, suggesting that our traditional rules of field theory are hiding a common underlying structure.
- This may revolutionise our understanding of field theories but also has practical applications in being able to do calculations that were previously impossible.

Chapter 10

The Square Root of a Black Hole and Other Curiosities

10.1 How General Is the Double Copy?

There were a lot of grand words in the previous chapter, in which I tried to convey quite how exciting I think the double copy is. But let us not get ahead of ourselves: at the moment, all this amounts to is a procedure for generating very particular and complicated quantities (scattering amplitudes) in gravity theories, based on similar results in other theories. Even given the string theory explanation of where this may come from, how do we know that this is not merely some sort of coincidence that only happens to be useful for these particular quantities but not much else? If this is the case, it in no way diminishes the importance of the double copy as a useful tool for calculations in quantum field theory. But it would be a shame, given that it would be much more profound for our understanding of the universe if the double copy turns out to be a lot more general.

Another way of thinking about this question is to look again at Figure 9.5, and ask the following: how generally are we to interpret this picture? The minimum we know is that this scheme only applies to scattering amplitudes. But what if we choose to interpret this picture as being really true for the "complete" theories of gauge theory, gravity, etc? What does it even mean to ask such a question? Speaking from my own point of view, I first began thinking of such questions in about 2013, in collaboration with Ricardo Monteiro and Donal O'Connell. We first met at an annual conference devoted to the study of scattering amplitudes and related topics that was held that year in a bizarre castle 50 km south of Munich. We asked ourselves the following question: are there other kinds of quantities, besides scattering amplitudes, that we can apply the double copy to? After all, in the previous chapters of this book, we have seen that there are many interesting objects in both gauge and gravity theories, such as charged particles that can move about and generate fields (gauge theory) and black holes or expanding universes (gravity). What all of these things have in common is that the fields they generate are exact solutions of the equations describing the relevant field in each case. For example, one of the possible solutions of Yang–Mills theory corresponds to a stationary particle with colour charge that creates a gluon field filling all space. Likewise, in gravity, we know that the Schwarzschild solution that describes the most basic kind of black hole arises from solving Einstein's equations of General Relativity. If we could somehow show that a solution of gauge theory could be "double copied" to gravity, it would be very nice for the following reasons:

- Showing that the double copy works for exact solutions of different theories reveals that it is much more general than had been previously thought.
- As we saw previously, scattering amplitudes are only ever calculated approximately. However, certain black hole solutions do not involve any approximations, so this gives us a much stronger statement about what the double copy is.
- Solutions of gauge theory are typically easier to find than solutions of gravity theories. Thus, if we are able to double copy exact solutions of gauge theory, this might give us new ways to generate solutions of General Relativity and related theories.

We decided fairly quickly to focus on the well-known Schwarzschild black hole solution of General Relativity, described in Section 7.3.2. After all, black holes are glamorous, at least by physics standards. The Schwarzschild solution is also the simplest type of black hole, so a natural place to start in trying to find a double copy interpretation. Our task was to find some sort of field solution in a gauge theory, and a procedure for double copying this, so that when we applied this to the relevant equations, we would get our desired black hole.

As so often happens in science, we spent well over a year trying different things, most of which were complete nonsense. Some of the things did make sense, or at least we thought they did, but they just didn't seem to go anywhere or tell us anything. Finally, we stumbled upon a particular way of writing the Schwarzschild black hole that is in suitably dusty old books if you know where to look, and which very much looks like something has been copied. It is called the *Kerr–Schild form* of the black hole solution and was first discovered in the 1960s, using methods that were then used to discover the rotating black hole solutions discussed in Chapter 7. Finding this form of the solution allowed us to "guess" the form of a gauge field that could be the single copy of the Schwarzschild black hole. That is, it would be the solution of gauge theory such that, if it was double copied, it would turn into a black hole. It is a straightforward exercise to plug this gauge field into the equations of Yang–Mills theory and, lo and behold, it solves them. It also turned out to be a much simpler solution than we had thought it would be.

To understand the single copy of a black hole, let me tell you that the easiest way to make a black hole in gravity is to pack a lot of mass into a small volume. The most extreme case is to take a completely point-like particle (i.e. of zero size) but demand that it has a finite mass. If we plug this into the equations of General Relativity, and ask what solution we get, the answer is the Schwarzschild black hole. The graviton field is defined everywhere in spacetime, except at the very specific point where the particle is located. For the single copy, we are looking for a similar object that lives in a different theory (gauge theory) and thus generates a different type of field. It is nothing other than a stationary point-like charged particle that sits at a given point in space. Strictly speaking, we should think of this object as living in a gluon theory so that it corresponds to a colour charge. However, we also found that we can just as easily think of it as being a conventional electric charge that generates a simple electric field, without the full complications of gluons having to be present. Indeed, we have already seen the electric field generated by a charged particle in Figure 3.3, where we have also drawn attention to its similarity to Newton's theory of the gravitational field due to a massive particle (Figure 7.1). We now know that the double copy is sitting behind this relationship: the charged particle in electromagnetism is related by the double copy to a black hole in gravity, which reduces to Newton's equations for a massive particle as we move far enough away from the particle itself!

The fun didn't stop there. We then looked at rotating black holes, known as *Kerr black holes*. As first shown in the 1970s, the relevant graviton fields can be thought of as being generated by a rotating massive disk, which has a certain structure (i.e. the density of the disk is not the same everywhere). The single copy of this object turns out to be a rotating disk of charge, where some parts of the disk carry more charge than others, in exactly the same way that the massive disk in gravity does not have a uniform density. Of course, a charged object should create an electric field, so we can plot the electric field of the charged disk just as we did for charged particles in Chapter 3. This is shown in Figure 10.1 where, to avoid cluttering the pictures, I have chosen to indicate the direction of the field only but not the explicit arrows at each point which tell us the size of the field. The disk is taken to lie in the middle of each figure, such that we are looking at it sideways on. The disk stretches across the entire Figure 10.1(a), after which we progressively zoom out in Figures 10.1(b) and 10.1(c). What we can see is that the electric field has some interesting structure very near to the disk, which is ultimately due to how the charge is arranged on the disk itself. As we zoom out, the electric field becomes simpler and simpler, until it matches the field of a single charged particle, as shown in Figure 3.3(b). This makes sense: if we zoom out, we will not be able to see the different parts of the disk, and it must therefore look like a point-like object.

Given that the disk of charge is rotating, it will also generate a magnetic field. To see this, we can recall from Chapter 3 that moving charged particles create a magnetic field. A given part of the disk, as it moves around in a circle, will look like a charged particle. Thus, a magnetic field will indeed be generated. I have shown what this looks like (again, the direction only) in Figure 10.2. As in Figure 10.1, we are looking at the disk sideways on and then gradually zooming out. At large distances, the magnetic field looks remarkably like that of a bar magnet, as shown in Figure 3.5. It just so happens, for rather complicated reasons, that the magnetic field of charged particles moving in a circle is indeed like that of a bar magnet. Thus, for aficionados of electromagnetism, Figure 10.2(c) makes a lot of sense!

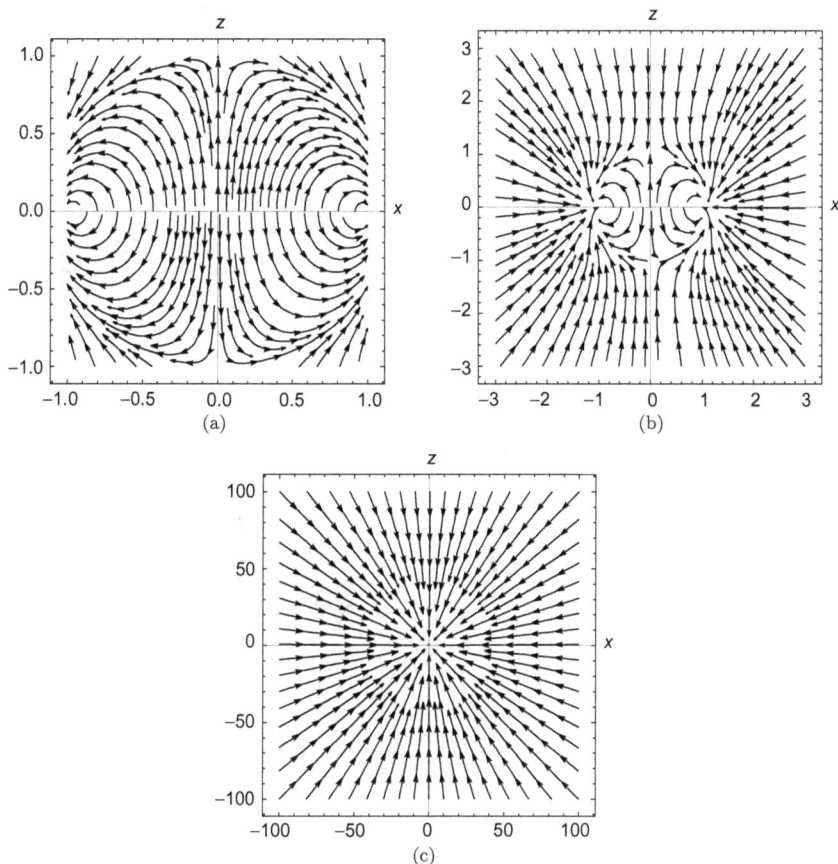

Fig. 10.1.   The single copy of the Kerr (rotating) black hole is a rotating disk of charge that generates the electric field shown, at: (a) short distance; (b) medium distance; (c) long distance.

These black hole solutions show that the scheme of Figure 9.5 can indeed be generalised to wider aspects of gauge and gravity theory. Crucial for this is the fact that one can also take a zeroth copy and find solutions of biadjoint scalar theory that correspond to the weird electric and magnetic fields we have found earlier. What's more, the Schwarzschild and Kerr black holes turn out to be special cases of the whole family of so-called *Kerr–Schild solutions*, which admittedly is still a rather special class of solutions of General Relativity. Still, given the nomenclature used earlier for taking the single copy, we

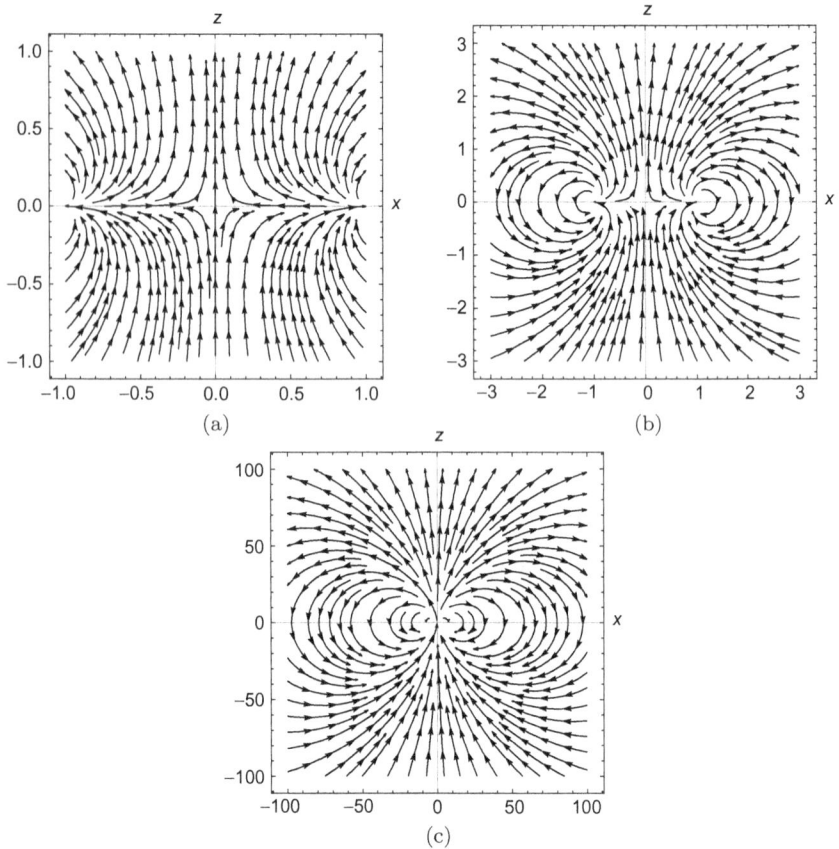

Fig. 10.2.   Similar to Figure 10.1 but showing the magnetic field instead.

can legitimately claim that we have been able to take *the square root of a black hole!*

10.2   Is This Really the Double Copy?

The earlier results clearly show that a procedure exists that can map solutions of equations corresponding to different theories. However, in order to properly be able to call this the double copy, we need to argue that it genuinely corresponds to the same thing as the double copy for scattering amplitudes that we discussed at length in the previous chapter. If we think a little about it, we can already see that there are distinct similarities. The double copy for black holes tells us that charged objects in gauge theory become massive objects in gravity. This replacement of charge by mass is a direct analogue of the replacement of colour by kinematic information in the amplitude double copy.

In the last ten years or so, we have understood in much more detail that the scattering amplitude double copy and the double copy for exact solutions of field equations are indeed facets of the same thing. One way of seeing this is that it is in fact possible to express solutions of field equations in terms of Feynman diagrams, even though these are not necessarily quantum. This was pioneered in a classic paper by Michael Duff in the 1970s, in which he showed that the Schwarzschild black hole can indeed be obtained from Feynman diagrams. These Feynman diagrams can be cast in such a way as to make the double copy show up between different theories and are precisely related to those used for scattering amplitudes. In recent years, Duff has developed an alternative approach to expressing the double copy for fields directly that offers useful complementary insights to the Kerr–Schild approach. Another approach to copying exact solutions is known as the *Weyl double copy* and was invented by Andrés Luna, Ricardo Monteiro, Isobel Nicholson, and Donal O'Connell in 2018. Further work has established that this is equivalent to the Kerr–Schild approach and is very precisely related to the amplitudes double copy. Thus, there can be no further doubt that they are the same thing.

What is nice about the double copy for exact solutions is that it starts to give us more intuition about what the double copy is doing, for the simple reason that charged particles and black holes are much less abstract to think about than scattering amplitudes. Thus, it is interesting to see whether there are related objects that can also be simply double copied. An interesting case is discussed in the following section.

10.3   Magnetic Monopoles

We have just now had occasion to remember that the magnetic field of a bar magnet is as shown in Figure 3.5. This makes us think about bar magnets, and what is special about them. As we saw in Figure 3.4, a magnet has two distinct ends, called a north pole and a south pole. When two north poles or two south poles come together, they repel. Two unlike poles, on the other hand, will attract. By comparing this situation with the electric charges we saw in Figure 3.3(c), whose electric field looks like the magnetic field of a bar magnet, it is tempting to conclude that if we cut a bar magnet in half, we will isolate two distinct types of "magnetic charge". However, if you buy a magnet (e.g. from a toy shop) and a hacksaw (probably not from a toy shop), you will be able to cut the magnet in half yourself. What you will find is that your two bits of magnet *both* act as smaller bar magnets. That is, by cutting the original magnet in two, you have somehow created a new north pole and a new south pole, as shown in Figure 10.3.

Fair enough, but what if you cut the magnet into smaller and smaller pieces? Surely at some point you will get an isolated north pole or south pole? No you will not. What is happening inside a bar magnet is a complicated quantum situation, but the upshot is that the magnetic behaviour comes from the electrons inside the magnet, such that each single electron itself looks like a combined north and south pole! Electrons are fundamental matter particles so that we cannot cut them in half. Thus, there is no way to isolate the north poles from the south poles.

This does not mean that we cannot ponder the existence of single magnetic charges (north and south) in nature. They have the hypothetical name of *magnetic monopoles* and have never been observed directly. Thus, in formulating Maxwell's equations, we leave out the possibility of magnetic monopoles, but we can easily modify the

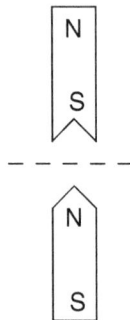

Fig. 10.3.   Cutting a bar magnet in half creates a new north and south pole.

equations should we ever find one. In fact, we want to find magnetic monopoles if we can. A particularly astute argument originally due to Dirac tells us that even if a single magnetic monopole exists anywhere in the universe, then we can immediately explain why electric charge in our universe only ever comes in fixed amounts. Quite why we've never seen the monopoles is a major puzzle, and there are international efforts to find them.

Whether or not magnetic monopoles exist, it is clearly a type of object that we can consider in a gauge theory. Thus, in keeping with the theme of this chapter, let us ask if one can "double copy" a magnetic monopole. That is, is there something in a gravity theory that a magnetic monopole turns into if we apply our copying rules? Indeed there is: a peculiar type of gravity object called a NUT charge, after some of the authors who first found it in the theory of General Relativity (Newman, Unti and Tamburino, although it was also found by Taub). For a black hole, the graviton field gets weaker as we go further away, such that spacetime approaches conventional flat space at large distances from the black hole. This is not true for a NUT charge, which generates a graviton field that looks like it is rotating at large distances and does not vanish. This is not the same as a rotating black hole and thus is a different kind of object.

While analogies have been drawn between magnetic monopoles and NUT charges before, the double copy makes this notion extremely precise. The magnetic monopole and the NUT charge live in *different* theories, but the equations describing the relevant fields are precisely related, throughout the whole of spacetime. Further work has linked this to scattering amplitudes, as well as showing how monopoles in arbitrary gauge theories can all be identified with gravitational NUT charges. Given that electric and magnetic charges are the only two distinct types of charges that can exist in electromagnetism, the identification of what their respective double copies are (mass and NUT charge) goes a long way to helping us interpret how the double copy may work more broadly.

10.4   Expanding Universes from the Double Copy

In Chapter 7, we saw that some of the most well-known solutions of General Relativity involve expanding universes. Also, that in our own universe, this expansion seems to be accelerating, due to the presence of dark energy. At late times in our universe, the matter that everything is made of will spread out and disperse, such that the universe can be approximated by a very large expanding spacetime that is filled with dark energy only. This is known as *de Sitter spacetime*, and it has been found that this spacetime also arises from a double copy. The relevant gauge theory solution is a uniform density of charge filling all of space, and this is entirely in keeping with our interpretation of the double copy for other solutions. We know that the double copy should replace charge with some sort of kinematic property, of which energy is indeed an example. Thus, the replacement of charge by energy is entirely natural from the point of view of the double copy. It seems to me that this aspect of the double copy may open up potential applications in the science of cosmology, which studies how our universe began and continues to evolve. To date, however, nobody has quite figured out what to do, although this situation may easily change in the coming years.

That exact solutions of gauge theory and gravity can be related by the double copy gives us hope that Figure 9.5 really does represent some deep connection between these theories. However, it remains the case that only very special solutions can be understood in this way. If anything, further research in this area has intimated that initial hopes of being able to copy any gauge theory solution to make a gravity one may only be possible for certain very special cases. That this is not a problem in practice arises from the fact that most gravity solutions of astrophysical interest can usually only be calculated approximately anyway. Thus, provided we know how to double copy approximate solutions of gauge theory, this is sufficient. We see the main application of this idea in the following chapter.

Summary

In this chapter, we have widened our interpretation of Figure 9.5 and seen that some of the most famous solutions of General Relativity can be put into this scheme. The main ideas are as follows:

- The different field theories of nature have equations that describe them. Solutions of these equations represent physical objects that produce these fields.

- We can formulate a double copy for solutions of gauge theory and gravity that allow us to obtain black hole solutions in gravity from counterparts in gauge theory.
- Electric and magnetic charges in gauge theory map to mass and so-called NUT charge, respectively.
- Expanding universes in gravity can also be obtained from the double copy.
- All of this tells us that the scheme of Figure 9.5 is a lot more general than being merely an accident of scattering amplitudes.

Chapter 11

The Double Copy in the Sky

As we saw in Chapter 7, General Relativity tells us that the force of gravity arises from the curvature of space and time, as if the latter are combined into some sort of stretchy fabric that can warp and bend according to whatever massive objects happen to be present. There can also be wave-like disturbances in the fabric itself that travel from one part of the universe to another. These are gravitational waves and can be described as the wave-like solutions of a graviton field obeying Einstein's equations. Quanta of these gravitational waves are the graviton particles we discussed in Chapters 8 and 9, whose scattering amplitudes we can obtain by double-copying gluon results according to the scheme of Figure 9.5. In this chapter, we see one of the main applications of this double copy in contemporary research, namely that it allows us to precisely understand the physics of gravitational waves.

11.1   Indirect Evidence for Gravitational Waves

Until relatively recently, we did not have direct evidence for the existence of gravitational waves, despite their being one of the earliest predictions of General Relativity. And until one has convincingly seen all the weird and wonderful consequences of a theory come true, one can never really quite trust it. That few people disputed the existence of gravitational waves was due to the fact that we did have very good indirect evidence, from a landmark experiment carried out over thirty years from 1974 onwards by Joseph Taylor and Joel Weisberg.

It consisted of watching a very distant pair of very heavy objects, where the first of these was a *neutron star*. These are the smallest and densest objects known in our universe, and they form when massive stars collapse. If you can imagine taking something much larger than our own Sun and then collapsing all of its material into a ball of radius 10 km or so, then you will have some idea of quite how dense a neutron star is. Were it any more massive, it would collapse to form a black hole.

Sometimes neutron stars can become highly magnetised, similar to how the Earth has a magnetic field. They can then end up emitting strong bursts of electromagnetic radiation from their north and south poles, which are easily detectable far away on the Earth. Given that neutron stars are typically rotating, we do not see a continuous burst of radiation, but a short pulse every time one of the poles of the neutron star crosses our line of sight, much like the light from a lighthouse. Thus, this kind of neutron star is called a *pulsar*, and such objects were first discovered by Jocelyn Bell (and her PhD supervisor Antony Hewish) in the late 1960s.

The pair of objects observed by Taylor and Weisberg consisted of a neutron star and pulsar orbiting each other, which collectively is known as a *binary pulsar*, where the word "binary" here implies that there are two objects. Given that the pulsar is orbiting the neutron star, we will see it moving towards and then away from us, as shown in Figure 11.1. By a clever trick, we can then measure how long it takes for the pulsar to do one complete orbit of the neutron star. In your everyday life, you may have noticed what happens when an ambulance travels towards and then away from you. As it crosses where you are and starts to move away, you will suddenly notice that

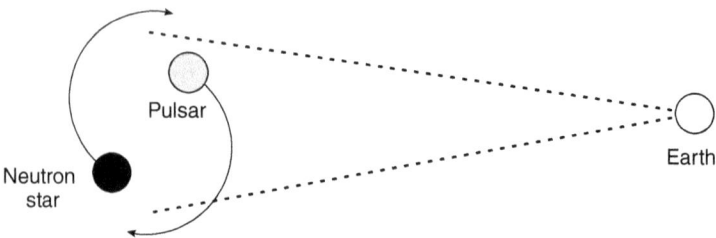

Fig. 11.1. In a binary pulsar, a neutron star orbits a pulsar, where the latter looks like a rotating lighthouse light. From the Earth, we will see the pulsar move backwards and forwards as it orbits.

its siren has a different sound (musically, it is a lower pitch). This tells us that the properties of sound waves get modified if objects start moving, and indeed the same happens for any kind of wave, including electromagnetic waves. What this means for the binary system in Figure 11.1 is that the rate at which the pulsar is seen to be flashing is *different*, depending on whether it is moving towards or away from us. Given that it moves towards and away once per orbit, physicists can then use this information to precisely calculate the time it takes for a whole orbit to take place. This is called the *period* of the orbit, and what Hulse and Weisberg did is measure this period on a regular basis, for over three decades!

For anyone to carry out such a repetitive task, they must have a pretty strong motivation. And indeed, General Relativity predicts that the period of a binary pulsar such as that in Figure 11.1 should gradually decrease over time, due to the fact that a pair of heavy objects orbiting each other should emit gravitational waves. These gravitational waves carry away energy and cause the orbiting objects to get closer to each other. Thus, it takes less time for a complete orbit to happen. In order to see this effect, we need some very heavy objects, such that the gravitational waves are significant. We also need to be able to clearly see one of the objects, which is why one of them being a brightly flashing pulsar is so useful. In Figure 11.2, I have shown a plot of the orbital period of the particular binary pulsar described earlier, as observed from 1974 onwards. We can see that this does indeed decrease, such that the orbits are getting slower and slower as time goes on. What's more, the solid line shows the prediction one obtains from General Relativity, i.e. assuming that gravitational waves indeed exist and are responsible for the effect. The fact that this line matches the data wonderfully is what convinced most people that gravitational waves are to be taken seriously!

11.2 Direct Evidence for Gravitational Waves

Having indirect evidence is one thing, but physicists are typically more ambitious than this. If something exists, then we would like to be able to see it directly. Also, there is a major motivation for wanting to observe gravitational waves. We have seen that they are generated by very heavy objects and have described in earlier chapters that

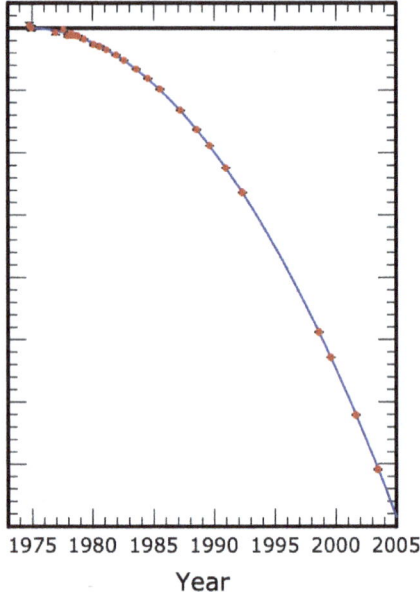

Fig. 11.2.   The orbital period of a particular binary pulsar, as observed over thirty years.

there may be large amounts of stuff in the universe, which we do not see directly as it does not interact very much with the Standard Model forces, e.g. dark matter. However, *all stuff*, dark or not, must interact gravitationally, as this only requires that something have a mass or an energy. Thus, being able to observe gravitational waves directly opens up a whole new window for viewing the universe we live in, allowing us to see some of its most unusual (and unusually violent) behaviours.

Gravitational waves were first observed directly in 2015 by the LIGO experiment. This experiment is based in the United States but, as often in modern physics, the team of scientists working on the experiment is large and based all around the world. There are over 1200 scientists from over 100 institutions in 18 different countries. The experiment itself consists of two independent detectors, each of which must detect tiny deviations in the structure of space-time, as a gravitational wave passes through the Earth. Each detector itself consists of a laser beam, which is split into two before being reflected from two mirrors in an L-shaped configuration, as shown

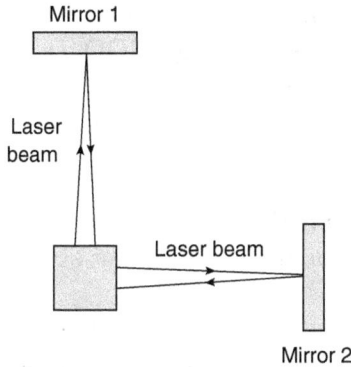

Fig. 11.3. Simplified picture of the LIGO experiment and related gravitational wave detectors. A central facility produces two laser beams that are reflected by two mirrors at right angles. The distances to the mirrors get distorted when a gravitational wave passes through, which we can measure by comparing the two reflected laser beams.

in Figure 11.3. The lengths to the two mirrors are equal in the absence of gravity. However, if a gravitational wave passes through the detector, the structure of spacetime will be warped. What this means in practice is that the distance to one of the mirrors might get slightly stretched, while the distance to the other is simultaneously squashed. We can measure this by comparing the two laser beams that get reflected from the mirrors, and this allows for a very precise measurement of the change in distance to each of the mirrors. In order to get this precision, however, the distance to the mirrors has to be very large, and in the LIGO experiment, this is around 4 km!

The detection of gravitational waves is hugely challenging, given that the deviations in spacetime due to the gravitational waves we typically see are very tiny indeed. The LIGO experiment is able to detect small distortions in the distance to each mirror which are 10,000 times smaller than the size of an atomic nucleus, making these comfortably the smallest measurements ever observed in a scientific experiment. This also explains why there are two independent detectors: by cross-checking the results, the scientists can make sure that they are not seeing spurious distortions of the mirrors, which are not due to gravitational waves.

As of November 2021, the LIGO experiment had seen 90 gravitational waves, produced from a variety of astrophysical sources, but each of them are analogous to the system shown in Figure 11.1. That is, a typical event observed by LIGO consists of two very heavy objects orbiting round each other. As they orbit, they radiate away energy in the form of gravitational waves, causing the objects to get closer and closer. Very often, one or both of the heavy objects are a black hole or neutron star. Then, as the objects get closer, they will eventually collide with each other, and merge to form a single object. Physicists then talk about three different stages of such events:

(i) *The inspiral phase*: This consists of the initial objects being well-separated, and orbiting round each other, emitting gravitational waves as they do so.
(ii) *The merger*: In this phase, the two orbiting objects come together and coalesce.
(iii) *The ringdown*: Having merged, the resulting single object wobbles a bit, before settling down to make a smooth black hole or neutron star.

The calculations needed to describe such processes are some of the most difficult in all of science, where much of this complication is due to the complex nature of General Relativity. Typically, describing the merger phase requires supercomputers, which crunch the full theory of General Relativity in order to try to describe the very strong gravitational fields taking place as the black holes or neutron stars merge. However, for the inspiral and ringdown phases, approximate methods may be used, and it is here where the double copy can play an immediate role.

11.3   Why Is the Double Copy Useful?

If we are far away from the colliding objects in a black hole or neutron star merger, then we will not immediately notice their structure. Thus, they will look like particles following line-like trajectories, just as our colliding particles did in Figure 8.2. It follows that we should be able to use similar methods in describing colliding black holes and neutron stars, as we use in quantum field theory for describing particle interactions. We saw that certain quantities called *scattering amplitudes* were the relevant thing in that context, and thus the following question arises: can our knowledge of scattering amplitudes be used to obtain results needed to understand gravitational wave experiments?

Over the past few years, there has been an intense effort to do just this, which has in turn brought together hundreds of people working at the interface between quantum field theory and astrophysics. In the prologue and intermission, I described one particular conference that has grown significantly in numbers, as more and more types of physicists have joined the fray. However, this is by now only one of many such meetings, and I can easily count at least three or four major international conferences per year, all of which aim to use quantum field theory methods to examine very applied and experimental questions in gravity. Pure mathematicians have also become interested, given that the kinds of calculation needed at the forefront of quantum field theory turn out to be intricately related to problems that have previously occupied mathematicians, for entirely different reasons. In my own career in science, I have never seen such a vibrant group of people who are willing to learn from each other and work together. However, this is not without its problems: science is now so specialist that even people who are working on the same underlying physics may be doing so using very different (mathematical) languages. Finding the common ground, and being able to translate between different ways of thinking, is what seems to occupy many of the discussions that go on at each meeting of the different subcultures. But this is also what makes the field so exciting: by abandoning traditional ways of thinking, new ones emerge.

Of the many approaches from the study of scattering amplitudes that can be used to address gravitational wave physics, the double copy is certainly one of them. As we learnt from Figure 9.5, the double copy allows us to obtain gravitational scattering amplitudes from the much simpler ones that arise in a gluon theory. Thus, new results for gravitational wave physics can be highly efficiently predicted, completely bypassing traditional methods of calculation, which are limited by the complexity of the algebra involved. Once we have scattering amplitudes in gravity, they can be used to increase the accuracy of our approximations for the gravitational waves emitted by binary systems, which allows us to understand the data coming from experiments more precisely. There are now a variety of approaches for taking measurable quantities in gravitational wave physics (e.g. how much the colliding objects get deflected, how much energy they radiate, and what the potential energy of the binary system is) and obtaining these directly from double-copied amplitudes. Further work has tried to show how the possible structure of neutron stars vs. black holes can be included and whether or not this is also amenable to being double-copied from a simpler theory.

So far, the use of the double copy in gravitational wave physics has been limited to the initial, inspiral, phase of a black hole or neutron star merger event. However, I have often wondered whether it may also have something to say about the ringdown phase. Preliminary evidence suggests that the "wobbles" of a black hole may in fact be related to wobbling charge distributions in a gauge theory, which would be a simpler system to analyse mathematically. The coming years are likely to see greatly increased effort in this area, with the double copy becoming a standard part of every physicist's calculational toolkit. It is also worth pondering whether there are other kinds of astrophysical processes that the double copy may be useful for. After all, every time that something happens requiring General Relativity – be it understanding the orbit of a heavy object, the expansion of the universe, or bouncing your latest social media post off a satellite – there is the potential to make the relevant calculations more precise. Doing so in coming years may increasingly involve using methods from quantum field theory, and hence the double copy and related correspondences.

Summary

In this chapter, we have described the sorts of astrophysical processes for which the double copy has already provided useful information. The key points are as follows:

- Gravitational waves are an unavoidable prediction of General Relativity, for which only indirect evidence existed until recently.
- They were directly observed for the first time in 2015 and open up a whole new way of observing our universe.
- A typical source of gravitational waves is a *binary system* of two heavy objects orbiting one another, before colliding and merging.
- The double copy is used to get more precise approximations for what is happening as the heavy objects orbit one another.

Chapter 12

Where Next?

12.1 The Web of Copiable Theories

So far we have seen that the double copy relates two types of theory that are directly relevant to nature. On the one hand, there are the so-called gauge theories that underly the Standard Model of Particle Physics. On the other hand, there is gravity, described by General Relativity and related theories. The last two chapters have shown that an increasing array of quantities can be related between these very different types of physical theories, suggesting some deep underlying connection between them. However, Figure 9.5 reveals that there is yet another type of theory – biadjoint scalar field theory – which is not directly relevant for nature as far as we know, yet which still seems to be related to theories that are relevant. This itself suggests that this may not be the whole story and that we should look for other theories that could be related by similar "copy relationships". There is by now a whole web of different types of theories that are known to be related in this way, where these theories have varying degrees of physical relevance for the world we apparently live in. At the time of writing, more than 20 different types of theories are known to fit into this picture. Also, there are intriguing cases in which information from different field theories can be used to make scattering amplitudes in a string theory, which goes beyond the original string theory motivation for the double itself and that we saw in Chapter 9. Many of these instances have been met with surprise as they have occurred, and it again all adds up to the suggestion that there is a vast underlying structure to these theories that our traditional ways of thinking have been hiding up to now.

Given the plethora of different theories that obey the double copy, you may be wondering whether there are any theories that cannot be obtained by copying other theories. The current state of play may perhaps be summarised by saying that there are certainly examples of theories that we do not know how to get by copying other theories. However, that does not rule out in such cases that there may be some new trick or way of thinking about things that might make a double copy possible in the future. What we seem to be missing is a systematic way of saying which theories are copiable into other theories and, consequently, whether the set of "copiable" theories is somehow picked out or special for some crucial mathematical reason.

The ideas of this section form just one of the motivations that has led researchers to try and find new ways of thinking about the foundations of quantum field theory and, in particular, to find new ways of representing scattering amplitudes. An especially fruitful one in a double copy context is described in the following section.

12.2   The CHY Equations

Let us look again at Figure 8.2. This shows a number of incoming particles, which then interact and produce a number of outgoing particles. Now imagine that we surround the point of interaction by the surface of a ball, which captures all of the particles as they emerge. Each of the outgoing particles will puncture the ball as it passes through it, and thus we can think of the properties of the interaction as being entirely determined by the positions of a set of points on a ball. We show this idea for the less simple case of four outgoing particles in Figure 12.1.

We can formalise this idea by writing a set of equations that express scattering amplitudes in terms of points on a sphere. Instead of traditional Feynman rules and diagrams, we instead get a much more abstract set of formulae for what a scattering amplitude looks like, involving summing over all possible places where the points might be, and also taking care to eliminate the freedom we have to choose how we talk about the sphere itself. This was first done by Freddy Cachazo, Song He, and Ellis Ye Yuan in 2014, and hence the

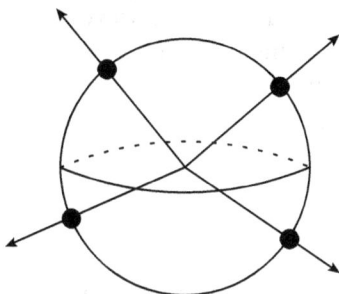

Fig. 12.1. Particles emerging from an interaction can be taken to puncture a ball-shaped surface surrounding the interaction point.

resulting equations have been known as the *CHY equations*. Interestingly, this description can be used for scattering amplitudes in different types of theories. When we examine what the formulae look like for different theories, we see that there is a natural double-copy-like structure, such that formulae in one theory can be obtained by recycling results from one or more other theories. Indeed, the CHY equations allowed people to discover some of the more general double copy correspondences we discussed earlier, for the first time.

There is another attractive feature of the CHY equations that concerns the string theory that we met in Chapter 9. There, we saw that interacting strings form worldsheets that are the analogue of the worldlines formed by interacting particles. In traditional mathematical approaches for calculating closed string scattering amplitudes for processes such as those depicted in Figure 9.3(b), it turns out that we are allowed to stretch and deform the worldsheet as it appears in our calculations so that it simply becomes a sphere. Each of the "holes" in Figure 9.3(b), associated with the incoming or outgoing closed strings, then gets mapped to a point on the sphere. This is highly reminiscent of Figure 12.1 and suggests that the CHY equations amount to writing field theory amplitudes in a very string-like way. More colloquially, the CHY equations amount to a field theory that is wearing a string vest!

That this idea can be taken further has been shown by Lionel Mason and David Skinner. They considered a highly exotic type of string theory called *ambitwistor string theory*. This is obtained from conventional string theory by replacing our usual spacetime with a very abstract mathematical space called *ambitwistor space*, whose details shall not trouble us here. However, the effect of this is that all of the complicated things that make strings different to particles disappear, leaving only the particles that should appear in a field theory. Follow-up work has tried to establish whether ambitwistor string theory can be obtained as some sort of extreme situation in conventional string theory, bolstering the idea that the double copy may ultimately have some sort of string-theoretic explanation. Might, for example, the copiable field theories all arise from certain (ambitwistor) string theories? What, then, are we to make of theories that cannot be furnished with such an explanation?

12.3   The Amplituhedron

The idea of representing scattering amplitudes in increasingly abstract ways has perhaps been taken furthest in a set of ideas that is typically referred to as the *amplituhedron*. This body of work was initiated by Nima Arkani-Hamed and Jaroslav Trnka in 2013, who found a remarkable property of certain scattering amplitudes in the $\mathcal{N} = 4$ Super-Yang–Mills theory that we saw in Chapter 9. They found that there are certain types of abstract mathematical spaces, such that one may consider well-defined shapes in them. Then, each of these shapes can be identified with a scattering amplitude in $\mathcal{N} = 4$ Super-Yang–Mills theory, in that known properties of amplitudes emerge from the pure geometry of the abstract shapes. Each such shape is known as an amplituhedron, and it remains unknown whether this description can be extended to theories that are more closely related to experiments (e.g. the pure Yang–Mills theory that describes interacting gluons only). More recent work by Trnka has discussed whether there is a corresponding *gravituhedron* idea for gravity theories, in which case there may well be an interplay with the double copy.

What makes the amplituhedron ideas so compelling is that they genuinely do turn our traditional notions of physics upside-down. For example, when we discussed the motivation for Special Relativity in earlier chapters, we discussed the need for our theories of physics to be *local* so that physical processes can only influence things which are near to them. A failure to require this results in crazy behaviour such as objects on Earth suddenly influencing things on the either side of the universe, and a major historical motivation for field theory was that it circumvents such issues. When we discussed quantum theories, we saw that we can only ever calculate the likelihood or probability for something to happen, rather than being able to predict with certainty what will occur. Implicit in this is that there must be a 100% chance of *something* happening, if we are not fussed about which option the universe ultimately picks. This imposes some sort of constraint on our quantum theories, which is given the fancy name of *unitarity*. In our traditional way of thinking about quantum theories, this property is built into the equations in a very transparent way, such that we don't really have to worry about whether or not it is true. However, in the amplituhedron way of thinking, both locality and unitarity are not at all obvious but instead emerge from a more mysterious underlying description of quantum field theory. This has raised the intriguing prospect that, if we are to succeed in rewriting the foundations of physics to make previously hidden structures such as the double copy clear, then it may involve a paradigm shift in how we think about space and time. The things we took as primary motivations for our current physical understanding may turn out to be secondary consequences of something else. All of this has echoes of the turn of the twentieth century, in which deep-held notions about the fixed nature of space and time, and even the deterministic nature of science, had to be abandoned. It will be fascinating in the coming years to see whether our current textbook ways of thinking may get replaced.

12.4 The Double Copy on a Table Top?

If the double copy and its related correspondences apply to increasingly varied types of theories, it stands to reason that some of these theories may manifest themselves in physical situations we have not yet considered. A major application of field theories, for example, is not in particle physics or gravity at all but in order to describe the behaviour of solid materials. Recall that the most basic idea of quantum field theory is that wave-like vibrations of fields get "quantised", where the little lumps or quanta of the field can then be interpreted as particles. This happens in solids all the time. For example, metals contain a regular lattice of atoms, through which electrons can move about. The lattice itself can vibrate in different ways, where these vibrations can be thought of as waves that travel throughout the metal. Quantum theory then tells us that these lattice waves get quantised, giving rise to a certain type of particle which is called a *phonon*. The phonons can be described by a quantum field theory, at least in the approximation that the wavelength of the "vibration waves" is very large compared with the distance between the atoms in the lattice.

Similar particle-like entities arise in all kinds of different materials, and one of the central questions in the study of solids – nowadays part of the subfield of *condensed matter physics* – is to identify the appropriate particles to talk about in any given situation and the relevant field theory if needed. When we construct equations for field theories, there are only so many possibilities that we can write down. Thus, it seems likely to me that some of the field theories that emerge when we study the double copy may well end up being used to describe condensed matter systems. What might this be useful for? Might there, for example, be calculations needed for understanding certain materials that become easier once we can copy them across from a simpler theory? Or are there new ways of thinking about condensed matter systems that come from reformulating the laws of quantum field theory? I suspect too that the biadjoint scalar field theory that we saw in Figure 9.5 may also correspond to some kind of exotic behaviour in a solid or perhaps to an approximation of a more complete and physically well-behaved theory.

As well as condensed matter physics, the double copy might have an interesting interplay with the field of *optics*, which studies the properties of light, quantum, or otherwise. For example, it is known that one may combine different light beams to make a system that "simulates" the behaviour of gravitons. Could similar experiments be used to simulate the extra particles that accompany the graviton in the double copy (i.e. the axion and dilaton)? Once we have such combined light beams, could we pass them through interesting materials to mimic how gravitons and related particles travel through interesting spacetimes? Such experiments have been proposed to study possible analogues of black hole physics, and the attractive thing about them is that they would be small enough – and safe enough! – to fit on top of a table in an average-sized physics lab. While preliminary efforts have been made to enthuse researchers working in these different fields of physics, it is fair to say that there is not yet much interaction between researchers working on the double copy and those working in other subfields of physics. Thus, there is ample scope for exciting developments in the coming years.

12.5   Concluding Remarks

In the first half of this book, we reviewed why physics is the science that underlies how our universe works at its most fundamental level and also described the landscape of theories that underpin our current understanding. Each of these theories – from quantum field theory to General Relativity – involves seriously weird behaviour, from the point of view of our everyday lives. To describe the universe successfully is to be forced to abandon the "obvious" facts that time ticks at the same rate for everyone, that science allows us to definitely predict what will happen in any situation, and that the space we live in is passive and unaware of what is passing through it. None of these things are true, such that the lesson we learn very early on in physics is that we must abandon the prejudice of *common sense*, and let a combination of abstract reasoning, and detailed experiments, guide our thinking. The upshot is that all of our theories of physics are so-called *field theories*, in which the various force and matter particles that we observe all arise from mysterious fields filling all of space.

In the second half of this book, we have seen that the various theories of fundamental physics that have taken thousands of years to arrive at may be much more closely related than previously thought possible. By studying the so-called *scattering amplitudes* that describe how particles interact, we see that we can obtain these in gravity theories, by modifying results from a much simpler theory of gluons. This is called the *double copy* and has been subsequently shown to apply to many different kinds of quantity, including some of the most famous objects in all of physics, such as black holes. A possible explanation for where the double copy comes from may lie in a hypothetical theory called string theory that has the power to unify all of our current theories into a single overall description. However, this may not be a full explanation, and it also does not matter if string theory fails to be itself realised in nature. What's more, the double copy reveals new connections between field and string theories that were not known about before. Recent years have seen an increasing web of theories that are connected by double-copy-like correspondences, and it is not yet clear why certain theories are picked out over others in this web.

The double copy poses deep conceptual questions about our current understanding of physics: is there a new way of writing the foundations of quantum field theory that makes the double copy show up from the outset? Is there a special class of theories that are picked out as "copiable", and what is this trying to tell us? Are there new mathematical ways of thinking about space and time that are needed? Are there new types of theories that will show up, in addition to the field and string theories we already know about?

In addition to its potential for revolutionising our understanding of basic physics, the double copy has a practical angle. Can it be used to generate new results for use in astrophysics? Going beyond the example of gravitational wave science that has arisen so far? Are there experiments in condensed matter physics and/or optics that may shed light on the double copy? Or, going the other way round, are there theories we get from the double copy that may describe what is going on in certain materials?

Given the relative youth of study in this area, now is the perfect time for new people to enter the discussion. These may be established researchers working in other areas of science or mathematics, but they may also be the next generation of scientists, who have yet to decide to pursue this career. Most importantly, if you are high school student yourself, there is every possibility that you will be one of the people rewriting our laws of physics from the ground up, over the next few decades. It is only fair, then, that I explain a bit more about how you can become a scientist, given that this advice is often missing in books of this nature. If, on the other hand, you are the parent of a young person who is interested in science, please pass this book on, especially the following chapter. When it comes to working together to further the cause of science, the more the merrier!

Chapter 13

How to Become a Scientist

I had two major motivations in writing this book. The first was to get people interested in a subject very close to my heart, given that new ideas seem obviously needed in the next few years, in order to understand quite what is going on with the foundations of fundamental physics. The second was to advise people on how to become a scientist, given that they are not typically told how to do so in their youth. As we saw in Chapter 1, many – if not most – children are obsessed with how the world works and why it works the way it does. Show me a toddler who has learnt to speak, regardless of gender or social background, and you are exhibiting a natural scientist, who has yet to be corrupted by the scourge of common sense. By the latter, I mean those myriad theoretical prejudices that encumber our thinking as our lives progress and make it ever less likely that we will be able to think outside the narrow confines of what appears to be sensible and obvious! Unfortunately, one of those pieces of common sense that feels self-evident to many is that they are either not able to do science or mathematics or that it would be foolish to try. My hope is that demystifying the process of how one becomes a physicist for a living may show people that it is achievable after all. What's more, even those physics students that decide not to ultimately become physicists, very often end up in richly rewarding jobs that favour the particular training that physics provides. Examples of these jobs include – but are not limited to – teacher, banker, lawyer, data science/artificial intelligence developer, software engineer, journalist, media developer, healthcare professional, civil servant, and so on. What physics seems to equip people with is an ability to solve problems in a pragmatic and creative way and also to be able to explain complex ideas to a range of different types of people. The fact that physics ultimately underlies all of the other sciences, and that its research methods involve the most cutting-edge computing, means that it has something to say to everyone.

My own background in science started with a keen interest as a child, fostered by science kits as Christmas and birthday presents and books on the subject that I read intensely. I knew nothing about universities or how to go to them. Growing up in a rural area meant that the nearest such institutions were miles away, and I was also not surrounded by people who had been to university themselves, apart perhaps from my school teachers. I was lucky though to have very excellent teachers, who managed to convince us all why both physics and mathematics were important and that they needn't be seen as impossible things that only a talented few can ever do. Both my teachers and family encouraged me to try for a university place, and the experience completely changed my life! Every day in my work, I get to look at the most amazing ideas ever created by human beings and to discuss them with fellow scientists. I contribute, in my own small way, to a vast edifice of human knowledge that stretches back over thousands of years and gets us ever closer to understanding the biggest questions regarding the nature of our existence. I also get to teach physics to the current generation of university students, all of whom inspire me with their own new ways of thinking about a subject that I thought I knew. I have travelled the world to meet scientists in other places and have even lived in different countries in order to see how things are done there. If you yourself fancy a similar career, then what follows is a description of the path that worked for me. But the main message is that you should ask for help at every step of the way and be aware of quite how much help there is, should you wish to access it.

13.1   High School

Due to the varied readership of this book, I will refer to *high school* as being that education that typically takes place from the ages of 11–18, coinciding with *secondary school* in the UK. This is typically the age at which choices are made regarding which optional subjects to pursue for higher study, provided that you do not leave school at the earlier age of 16. For a physics path, you want to make sure that you are studying physics and mathematics at the highest levels you can in high school. Not all physicists have multiple higher qualifications (e.g. A Levels) in mathematics, but at least one mathematics course is typically expected if you wish to pursue physics at university. Many students take more sciences at high school level (my own choice was to take Chemistry alongside Physics, Mathematics, and Further Mathematics). Biology is a common choice as well as Chemistry, but it is also not unusual for students to take some other subject that interests them that may not be directly related to maths and/or physics.

Of course, both maths and physics have the reputation of being difficult subjects. One of the first things to realise in studying these subjects is that the main skills you need are enthusiasm and confidence. Unfortunately, confidence is one of the easiest things to lose if you are studying something that repeatedly seems to tell you how stupid you are. Take it from me: one of the defining features of being a working scientist is apparently not understanding anything that you are doing and being made to feel extremely stupid almost the entire time. None of this matters – if anything, it makes you feel even more excited in those rare moments when you actually *have* understood something and then get to see where it leads you next. The only difference between myself and younger readers is that I have twenty or so years more experience in coping with the difficulty of physics and having all sorts of systematic tricks for doing so. Enthusiasm is my main weapon against the soul-destroying loss of confidence that happens from time to time. If I get stuck in a research problem for weeks on end, I will try all sorts of things to get my confidence back. These include reading popular science books, talking to colleagues, and, most importantly, talking to the students at my university, whose own youthful enthusiasm reminds me of why I loved science in the first place. The confidence always comes back eventually, but only if you see it for what it is: a skill that needs repeated practice and that can be lost if we don't nurture it. Our confidence can also be fed by other people (and vice versa), provided we are courageous enough to be honest when we are feeling low.

Always ask your teachers for help if you are not understanding something. I have done a large amount of teaching in my career, for reasons that will become clear in the following. No teacher worth anything will ever mind being asked questions about their favourite subject, by someone who is keen to learn about it. As I tell my own university students, you may end up being so confused by what is in your physics or maths courses that you cannot even formulate a question to ask. This makes it *more* important to seek help, not *less*. Furthermore, these are the cases that, as a teacher, I enjoy the most. Trying to find out where a student's confusion is, is a fun research problem in itself. Also, by having to explain things in multiple ways until I find the "right" one for a particular student, I nearly always come away having understood the subject a little more myself!

As well as advice on the subject, your teachers can also give you careers advice and help with applications to university. Always speak to them about anything that might be bothering you, as it is their job to answer you or to point you to someone else who can.

The short summary of this section is as follows: *choose physics, choose maths, and seek help!*

13.2   Undergraduate and Master's Degrees

The standard route to being a physicist involves pursuing a degree at a university, where there are different types of degrees. Broadly speaking, these are split into the following types:

(i) an *undergraduate degree* lasting three or four years. In the UK, this typically leads to a qualification called a *Bachelor of Science (BSc)* or *Bachelor of Arts (BA)*.

(ii) a *master's degree*, lasting one or two years, in which you can specialise further in a particular area of physics.

Note that it is possible to sign up for both of these at the same time. This is called an *integrated master's degree* and will lead, upon graduation, to two distinct degree certificates being awarded, for the undergraduate and master's degrees, respectively. Candidates can also leave in the middle of the master's degree and graduate with the undergraduate degree they have already obtained as part of the course. As an example, my own undergraduate and master's degrees lasted three and one year, respectively, and were integrated. At my time of studying in the UK, it was not possible to leave after three years and get state funding for a separate master's, which made an integrated master's a very common choice. Nowadays, however, one can indeed qualify for funding for a master's course somewhere else, which has led to much more competition between universities.

Undergraduate degrees may be in straight Physics or may already consist of more specialist pathways, such as Physics with Data Science, Theoretical Physics, Astrophysics, and so on. You may also see degrees that combine Physics with another subject, e.g. Physics with Chemistry, which will include roughly equal content from both disciplines. At the time of writing, you can search for undergraduate courses featuring Physics in the UK by going to the website of the *Universities and Colleges Admissions Service (UCAS)*.[1] You will find individual course pages from different universities, which will tell you important information, such as entrance requirements, employment rates, and student satisfaction. By going to specific university webpages (e.g. by Googling a university name followed by "Physics"), you can find out more about the structure of what you will learn in a particular course. One other thing to look out for is whether a given degree is *accredited*. This means that an external organisation – the UK Institute of Physics – has approved the degree as having rigorous standards and providing an excellent training for being a physicist.[2]

Your teachers should be able to help you in applying to university, writing personal statements, and so on. Be honest about your ambitions to your teachers, and feel free to ask them about their experiences and how they felt about university themselves. You may also want to attend open days at universities and, even if you don't, you should always feel free to contact people whose details you might find on university webpages. Often, they will be able to put you in touch with admission staff or students, who should be more than happy to talk to you about how to apply and/or what it's like to study in a given place. If they are not prepared to do this, what does that tell you about whether you would like to study there?

---

[1]The website can be found at https://digital.ucas.com/search.

[2]The Institute of Physics itself is a wonderful organisation for promoting the subject and supporting students, teachers, and researchers at all levels. Its webpage can be found at https://www.iop.org.

Let's now assume that you have plucked up the courage to apply to a university and are sitting there on day one of a physics degree, having possibly travelled a long way from home. What actually happens, and how will you learn? The first thing you will notice is that there are many more people around you than there ever were in a school class. Commensurate with this is the fact that you are expected to learn much more independently at university than at school. In particular, you will be taking much more control of your own learning. This is ultimately necessary to become a scientist – the aim of a university is after all to produce independent thinkers. However, the transition to university life can be very disorienting, particularly for those students who do not come from a middle-class background. This hits you at the same time as the fact that you are learning a hugely difficult subject! In your first year, you will see material that seems familiar from your schooldays, but you will be doing it in much more detail, and deriving things from scratch rather than relying on ready formulae and equations. You will also start to see the exotic ideas behind relativity and quantum mechanics, if you have not done so already. If this does not make your head ache at times, then you are in a tiny minority. Thus, it is only normal in some circumstances for one's confidence to be affected and/or to suffer from "imposter syndrome", namely the feeling that everyone else belongs but that you do not. Having seen thousands of students in my career, I can confidently tell you that this is all perfectly normal and that the solution is once again to seek help and to talk to people. Be honest about how you are thinking and feeling, both to your fellow students and to staff. I strongly feel that us scientists do not talk enough about all of the things happening besides the physics that affect our ability to learn. For undergraduates, the reason for this is almost always that students feel they cannot talk about non-physics things with a member of physics staff. You can and should talk to any staff member that you encounter, and you will usually be appointed a dedicated staff member whose job it is to look after your well-being. Trust me, they will have seen almost all crises of confidence before, including their own! And let me give you a simple trick that has always worked for me regarding imposter syndrome: I simply remind myself that I *am* an imposter given the background I came from but that it hasn't prevented me (nor anyone else in similar circumstances) from doing science and loving their job!

Depending on your degree and institution, you will do a mixture of physics-related activities, such as learning physics theories from lectures, practising your understanding of these by doing exercises and problems, discussing with staff and fellow students in small groups, and performing lab experiments and analysing the results, both to learn physics theory and to gain important experimental skills. Physics courses increasingly include dedicated computing courses. It is virtually impossible to do research in any field of physics nowadays without doing some programming, and the topics in this book are no exception. Finally, as you progress up the degree, you will start to undertake independent project work, in collaboration with a member of staff. This is your first taste of doing actual scientific research, which is a very different skill to learning physics theories or passing exams. I have often seen students who struggled with lecture-based work really come into their own when it comes to project work. And I always enjoy reading their dissertations (lengthy reports summarising the project)!

When it comes to which subjects you will learn at different stages in your degree, this can vary in different courses. Roughly speaking, your knowledge of relativity and quantum mechanics starts in year 1, as well as a detailed understanding of the theory of electromagnetism. Nuclear and/or particle physics will follow in years two and three, as will the theory of solids (which requires a lot of quantum mechanics to get right). Some particularly important subjects deal with what happens when you have seriously large numbers of particles in one place: thermodynamics and statistical mechanics. People tend to hate these when they first see them (as did I), only realising later on how these subjects underpin almost all of physics! Dedicated maths courses from year 1 onwards will equip you with the ability to understand the language of modern physics, and if you are learning astrophysics, this will have its own timetable of important ideas that you will learn in different years.

In almost all universities, there will be optional courses in later years, allowing you to specialise in certain subareas. This is when you are likely to encounter General Relativity for the first time. As discussed earlier, your master's course may see you moving to another university, or you may choose to stay at your existing one. In either case, most master's courses usually include an extended piece of project work and more specialised courses on your chosen area of physics – such as quantum field theory. It is much more common at master's level to see courses which are unique to the research expertise of a particular university, although this can sometimes happen at lower levels too.

By the end of a degree, you will have seen an enormous amount of physics, be equipped with a wide range of skills that make you employable in a wide range of high-powered jobs, and also gotten your first taste of what scientific research is like. If you still prefer doing the latter for a living, then a PhD is your natural next step, which is to be discussed in the following. But what happens if you have decided that physics is not for you, after all? There is absolutely nothing wrong with this, nor wrong with you. In fact, it is incredibly common for people to realise that their university years have taught them what they don't like, rather than what they do. All experiences, however, are useful in your life, if thought about in the right way. And the skills you learn in a physics degree will set you up for life in many different jobs. Talk to staff at your chosen university, and they will be more than happy to help you explore your careers options, which may include passing you on to dedicated career experts outside the physics department.

13.3   The PhD

Your first degrees (undergraduate plus master's) can take four or more years. After this, budding scientists will complete a special type of degree called a *Doctor of Philosophy (PhD)*. In some places (e.g. Oxford), this may be called a DPhil instead, which means the same thing. A lot of nonsense is thought and written about PhDs, such as the fact that they are impossibly difficult and/or that you have to be a genius to do one. Neither of these things is true, and the best way to think about a PhD is that it is an apprenticeship. For three or more years, you will be stationed with a working scientist (your *PhD supervisor*), who will mentor you to conduct an extended piece of scientific research work. They may identify a suitable problem and initial reading material, assist with the project direction, assist with performing calculations, coding, or other such tasks, and provide more general advice and confidence-boosting as needed. You may have other members in your supervisory team (e.g. a second supervisor or graduate tutor) and also be part of a wider research group of people working in a similar area. The PhD ends when you have accrued sufficient material to write a PhD thesis, which is a lengthy document of around 200 pages (perhaps shorter or longer, as required). In my own field of theoretical physics, the thesis should usually contain enough material for three scientific research papers, as they are published in scientific journals. Indeed, it is very common for PhD candidates in my field to have already published their work while the PhD progresses, such that the thesis has to tie together the results and explain them in a wider context. The PhD is assessed by one of the most nerve-inducing examinations ever faced in a scientific career: the PhD viva.

In the UK, the viva is of indefinite length and consists of two examiners – from both outside and inside the candidate's university – who have read the thesis and ask the candidate to "defend it". Provided the candidate has actually done the work and written the thesis (n.b. if no serious fraud is involved), the viva should take the form of an extended and relaxed chat, where the two examiners clarify points in the thesis and learn from the student about what their results mean. Unlike most other exams, the expert in the room is very much the student themselves. My own experience of a PhD viva is that I was needlessly terrified for months before it occurred, only to find that my examiners were two wonderful scientists who seemed genuinely interested in what I had done and how it might relate to their own work. What I tell my own PhD students who are preparing for this rite of passage is that it is very rare indeed that an entire afternoon or morning is devoted to themselves and their own work! It is a chance to show off everything they have achieved and to get to know a leading scientist in their field who may well help their career moving forwards. I have learnt a lot over the years from examining PhD vivas and always look forward to seeing what the students I have examined get up to next.

Seen as an apprenticeship followed by a chat, a PhD is perhaps less intimidating than it might first appear. It can, however, be another of those moments of low confidence in a scientific career. The main reason for this is that most of our studies before this point involve conventional exam-based assessments, which apparently rely on there being a right and a wrong answer. Without us even noticing, this constantly gives us feedback about how we are doing, such that we feel incredibly uncertain if this feedback is suddenly removed. In a PhD, the focus shifts fully for the first time to discovering new things that nobody else knows but the people discovering them: the student, supervisor, and additional collaborators if they exist. One way I explain this to my own students is that they are transitioning from asking "What is the answer?" to "What is the question?". Knowing that this uncertainty is coming and perfectly normal can be a great help. If you wanting to go down the PhD route, know again that you can talk to your supervisor about *anything*, and they will have their own stories about how they coped with becoming a fully fledged researcher.

13.4   Postdoctoral Positions

After the PhD, many people decide that they have had enough of academia and are very well qualified to take up a job in industry or education. There are certain high-powered jobs (e.g. in finance or computer science) that require a PhD, regardless of whether the subject matter directly aligns with the job at hand. What employers are looking for is extended experience in time-managing a lengthy project, which is much more complex than the shorter projects that are carried out in undergraduate or master's degrees. Thus, you should never worry that doing a PhD has been a waste of time, if you decide to leave academic research afterwards.

   For those that want to carry on being a researcher in fundamental physics, the next thing is to apply for *postdoctoral positions* ("postdocs"). These are highly competitive, and thus you need to apply for many of them and not be too fussy about where you end up. In my own case, I applied to roughly forty positions throughout Europe, but this can easily rise to a hundred or more if you include the United States and/or Asia (which I excluded at the time for personal reasons). At this stage, it is very common to move to another country, which can itself be fantastic and a great way to see the world. However, there is no denying that this aspect of the job gets increasingly difficult if you then have to move around every two years or so until something more permanent turns up. The fact that one has already done at least seven years of training before even getting to the postdoc stage exacerbates this problem, and recent years have seen an increasing focus on supporting early stage researchers so that they can make informed choices and also be highly employable whatever comes next.

The aim of your postdoctoral years is to build up an independent research profile, by publishing scientific papers in a particular area or range of areas. This by no means entails that you have to work alone: almost all scientific papers (including in theoretical physics) involve teams of people working together. However, to get to the next stage of your career, you usually need to have demonstrated some kind of leadership, meaning that you are strongly associated with a particular programme of research that you are heavily influencing.

13.5   The Permanent Job

The holy grail for any postdoctoral researcher is to land a permanent job. Very very few of these are research-only and are based in government research institutes in various countries. Instead, the vast majority of permanent positions are in universities, where staff are expected to carry out and publish research while also teaching the next generation of students. If you study a physics degree at a UK university, most of your lecturers will be working scientists in this manner, allowing you privileged access to the latest scientific developments. As a lecturer myself, I relish both my teaching and research roles. They are not at all the separate activities that some people think they are, and there is no doubt that teaching students directly informs my research and vice versa.

As well as carrying out teaching and research, the life of a scientist also involves being actively involved in the management of a university department or its courses, or even aspects of the wider university. It can also include performing so-called *outreach* activities, such as giving talks to the public or writing books such as these. The wide range of activities associated with modern day academia means that universities are increasingly looking for all-round people with a wide range of skills, rather than being narrowly focused researchers who are incapable of doing anything else. This is particularly true in the UK, in which various external government pressures regarding teaching quality have arisen in recent years. Anyone and everyone is welcome, provided they are motivated to rise to the challenge and can convey their passion and enthusiasm to others. If this sounds like you, then please consider taking up science for a living. I am genuinely excited to see what you go on to achieve!

Epilogue

It has often been said in recent years that fundamental physics is in a crisis. The twentieth century destroyed existing paradigms and gave us an astonishingly complete list of theories with which to understand our universe: quantum field theories for all matter particles and three of the four fundamental forces in nature; General Relativity for gravity, which also describes the very beginning of the universe itself. Literally the only thing wrong with these theories is that they cannot be correct! The Standard Model of particle physics is full of mysteries and cannot explain dark matter or dark energy. General Relativity itself breaks down in those very places that we would like to most use it, namely the centres of black holes and the Big Bang itself. None of this should stop us admiring the amazing insights and structures that these theories give us, but we can – and should – hope for more.

At the time of writing, our species' fine tradition of particle accelerator experiments has spectacularly confirmed the role that the Standard Model plays in our universe, including finding the seriously abstract and elusive Higgs boson. Having achieved its design goal, however, it has provided scant hints of what lies beyond our current theories, such that it is not at all clear where the next big leap in physics will come from. Experiments in astrophysics or cosmology (the study of the early universe) may provide hints, as might the emerging new science of gravitational waves. What is certain, however, is that new calculational tools and methods will always be needed, in order to compare our theories with experiment and understand the consequences when they break down.

Our hints in recent years that the foundations of physics need to be rewritten have been described in this book and tell us that the next few years may be crucial in defining the answers to the big questions facing our existence. The history of this subject tells me not to be foolish enough to guess what will happen next or which groups of people will be responsible for changing our thinking. My only request – as simple as it is humble – is that you consider joining the team.

Index

www.ingramcontent.com/pod-product-compliance
Lightning Source LLC
Chambersburg PA
CBHW061247220326
41599CB00028B/5563